Paul W. Holland

CHARLES GRIFFIN & COMPANY LIMITED
16 Pembridge Road, London W11 3HL, U.K.

First published 1987

Published in the USA by Oxford University Press, New York, N.Y. 10016
ISBN 0–19–520615–0

British Library Cataloguing in Publication Data

Bartholomew, D.J.
Latent variable models and factor analysis.
—(Griffins statistical monographs &
courses; 40)
1. Latent variables
I. Title
511′.8 QA278.6

ISBN 0-85264-280-6

Typeset in Northern Ireland by
The Universities Press (Belfast) Ltd.
Printed & bound in Great Britain by
Redwood Burn Limited, Trowbridge, Wilts

Latent Variable Models and Factor Analysis

D. J. Bartholomew

Professor of Statistics
London School of Economics and Political Science

Monograph No. 40
Series Editor:
ALAN STUART

CHARLES GRIFFIN & COMPANY LTD
London
OXFORD UNIVERSITY PRESS
New York

LATENT VARIABLE MODELS AND FACTOR ANALYSIS

Contents

Preface

The aim of this book is to give a unified treatment of statistical methods based on latent variable models. Among these, factor analysis is the best known and is the subject of a voluminous literature, but latent class and latent trait analysis are other important members of the same family.

The very unequal development of the field has dictated the somewhat unusual structure of the book. After a survey in Chapter 1 there follow two chapters covering the well-established topics of latent structure and factor analysis. The treatment here is along traditional lines but in a manner intended to bring out the common elements which are collected together in Chapter 4. This is the kernel of the book. It provides a retrospective rationale for the models and methods which precede it and a base from which to develop those that follow. These concern the much newer field where the data are categorical and the latent variables metrical. Hitherto there has been no definitive treatment of this area, so the subject is developed more fully over five chapters in which models, methods and applications are treated separately.

Factor analysis was invented by psychologists and has largely been developed by them for their own particular purposes. Their interests will continue to be a major influence but the technique is now used in fields as diverse as geology, sociology, chemistry and geography. As befits a statistician, I have therefore tried to take a broader view and this has materially influenced the choice of topics to be covered. Limits of space explain some omissions, but the introduction of others, such as image factor analysis, Q-factor analysis and much of the more esoteric theory of rotations, would have tended to obscure the essential simplicity of the basic theory. I hope that psychometricians will also benefit from this more sharply focused approach, even if they fail to find it completely satisfying.

From the statistical side, the book is in the tradition of Lawley and Maxwell's fundamental work *Factor Analysis as a Statistical Method.* However, it is much broader in coverage but less thorough in the treatment of mathematical detail. The numerical analysis side, in particular, is now taken care of in well-tried computer packages and so can give way to a fuller account of modelling. Indeed a principal tenet of the book is that the careful specification of a suitable probability model is the key to sound practice.

The book is written primarily for psychometricians and statisticians. For the former it gives a broad view of the state of the art. For statisticians it provides a treatment of a subject, which many of their clients will wish to use, in a manner which regards it as a serious part of statistical methodology. However, it would be particularly gratifying if other scientists, especially social scientists, were to find that it offered new ways of tackling their own research problems. For, above all, the approach provides a framework for thinking about that most basic element of science, namely measurement.

The applicability of most of the methods described depends heavily on the availability of the appropriate computer software. A note is included about programs currently available but this will quickly become out of date. Indeed, one of the objects of the book is to stimulate the production of new and better software.

I have used the material in courses at the graduate level and in this context, particularly, it is necessary to emphasize that latent variable methods are delicate tools. They are often seen as promising so much more than they are capable of delivering, with the result that enthusiasm is overtaken by disillusionment. A book can provide the foundation for sound practice but only good teaching and first-hand practical experience can complete the task.

In writing this book I have received much help from many sources. Mary Cahill has seen the typescript through several versions with speed and efficiency. My wife has helped with the bibliography and most of the other tasks required for the preparation of a typescript for the publisher. Alan Stuart suggested that I write the book, and various other past and present members of the Department of Statistical and Mathematical Sciences at the London School of Economics have given me the benefit of their experience. Brian Shea, in particular, has been closely involved at all stages. His work on the computing side was supported by ESRC grant H00232012. Pat Moran of the Australian National University read much of the book in typescript and his suggestions were the means, direct and indirect, of many improvements. The writing has had to be fitted into a heavy programme of other work and the latter stages were greatly accelerated by the opportunity to spend most of March and April 1986 at the University of Melbourne. I am very grateful to the London School of Economics for providing the time, and to the Department of Psychology and Queen's College, Melbourne, for the warmth of their hospitality.

London
1987

DAVID BARTHOLOMEW

Notation

1. Matrices and vectors are denoted by bold face type. Lower-case symbols are usually used for vectors and capitals for matrices. The dimensions of matrices are not made explicit in the notation but, for example, the phrase "q-vector" refers to a vector with q elements. The (i, j)th element of a matrix $\mathbf{A}$ may be denoted by $\{\mathbf{A}\}_{ij}$. A matrix inequality, such as $\mathbf{A} \geqslant \mathbf{B}$, means the same as

$$\{\mathbf{A}\}_{ij} \geqslant \{\mathbf{B}\}_{ij}$$

for all i and j.
$\{\mathbf{A}\}_i$ denotes the ith row of $\mathbf{A}$. $\mathbf{1}$ denotes a vector of 1's and $\mathbf{0}$ a vector or matrix of 0's. $D(\mathbf{x})$ is the dispersion (or variance–covariance) matrix of the variables $\mathbf{x}$; corr($\mathbf{x}$) is the correlation matrix; corr($\mathbf{x}, \mathbf{z}$) is a matrix whose (i, j)th element is the correlation coefficient between x_i and z_j.

2. $\{g_{ij}\}$ refers to the set of all elements g_{ij} obtained by letting the subscript(s) range over all values in the index set.

3. The goodness-of-fit statistic used is always the log-likelihood ratio statistic defined by

$$\chi^2 = -2\Sigma O_i \log O_i/E_i$$

where $\{O_i\}$ and $\{E_i\}$ are the observed and expected frequencies.

4. The symbol $\frown$ means "is distributed like" and $N_p(\boldsymbol{\mu}, \boldsymbol{\Sigma})$ refers to a p-variate multivariate normal random variable with mean vector $\boldsymbol{\mu}$ and dispersion matrix $\boldsymbol{\Sigma}$.

5. Summation signs usually have lower and upper limits, for example, $\sum_{i=1}^{p}$ but, for typographical convenience, may be abbreviated to $\sum_{i}$ or Σ.
A multiple integral is written either $\int_R \ldots d\mathbf{z}$, where R specifies the region of integration, or $\int \ldots \int \ldots d\mathbf{z}$ when the region may be specified by limits on the integral signs.

6. $\mathbf{x}$ always refers to a manifest variable and $\mathbf{y}$ and $\mathbf{z}$ to latent variables. No subscript will be attached if $\mathbf{y}$ or $\mathbf{z}$ has a single

element. Variables denoted by $\mathbf{z}$ usually have the range $(-\infty, +\infty)$ and subscripts may have *three* different meanings which it is important to distinguish:

(i) They may index latent variables (as with y) and in this usage j is used as the subscript.
(ii) They may index discrete values of a single latent variable and in this case t is used as the subscript.
(iii) They may index the value of a single latent variable assumed by the individual indexed by h which is then used as the subscript.

CHAPTER 1
Basic Ideas

1.1 The statistical problem

Statistical methods based on latent variable models play an important role in the analysis of multivariate data. They have arisen in response to practical needs in all the sciences but especially in psychology and other social sciences. They can be approached in an empirical, pragmatic way or theoretically, and we shall show how the need for them arises by taking each route in turn.

Large-scale statistical enquiries, such as social surveys, generate much more information than can be easily absorbed without drastic summarization. For example, the questionnaire used in a sample survey may have 50 or 100 questions and replies may be received from 1000 respondents. Elementary statistical methods help to summarize the data by looking at the frequency distributions of responses to individual questions or pairs of questions and by providing summary measures such as percentages and correlation coefficients. However, with so many variables it may still be difficult to see any pattern in their inter-relationships. The fact that our ability to visualize relationships is limited to two or three dimensions places us under strong pressure to reduce the dimensionality of the data in a manner which preserves as much of the structure as possible. The reasonableness of such a course is often evident from the fact that many "questions" overlap in the sense that they seem to be getting at the same thing. For example, one's views about the desirability of private health care and of tax levels for high earners might both be regarded as a reflection of a basic political position. Indeed many enquiries are designed to probe such basic attitudes from a variety of angles. The question is then one of how to condense the many variables with which we start into a much smaller number of indices with as little loss of information as possible. Latent variable models provide one way of doing this.

The matter may be put more formally as follows. We start with a *data matrix* containing observations on p variables for n sample members written

$$\mathbf{X} = \begin{bmatrix} x_{11} & x_{12} \dots x_{1p} \\ x_{21} & x_{22} \dots x_{2p} \\ \vdots & \vdots \quad\quad \vdots \\ x_{n1} & x_{n2} \dots x_{np} \end{bmatrix}$$

The x's may be measured variables or codes assigned to different answers to a question. The problem is to replace **X** by another matrix with a much reduced number of columns which is as near to **X** as possible in some sense. If the p variables were scores obtained on a series of tests of some motor ability one might feel that **X** could be summarized simply by summing the rows. The total score would then be a one-dimensional representation of the original p-dimensional data matrix. This procedure raises the questions of whether this is a reasonable summarization and what to do if it is not. Such questions can be answered by reference to a model.

The second approach to latent variable models is more theoretical and arises most naturally within a social science context. A cursory inspection of the literature of social research or public discussion in newspapers or on television will show that much of the discussion centres on entities which are handled as if they were measurable quantities but for which no measuring instrument exists. Business sentiment, for example, is spoken of as though it were a real variable changes in which affect share prices or the value of the currency. Yet business sentiment is a nebulous concept which may be regarded as a convenient shorthand for a whole complex of beliefs and attitudes. The same is true of quality of life, conservatism and general intelligence. It is virtually impossible to theorize about social phenomena without invoking such hypothetical variables. If such reasoning is to be expressed in the language of mathematics and thus made rigorous, some way must be found of representing such "quantities" by numbers. The statistician's problem is to establish a theoretical framework within which this can be done. In practice one chooses a variety of indicators which can be measured, such as answers to a set of yes/no questions, and then an attempt is made to extract what is common to them.

In both approaches we arrive at the point where a number of variables have to be summarized. The theoretical approach differs from the pragmatic in that in the former a pre-existing theory directs the search and provides some means of judging the plausibility of any measures which result. We have already spoken of hypothetical variables. The usual terminology is *latent* variables or *factors*. We prefer to speak of latent variables since this accurately conveys the idea of something underlying what is observed. However, there is an important distinction to be drawn. In some applications, especially in economics, a latent variable may be real in the sense that it could, in principle at least, be measured. For example, personal wealth is reasonably well-defined concept which could be expressed in monetary terms but in practice we may not be able or willing to measure it.

Nevertheless we may wish to include it as an explanatory variable in economic models and therefore there is a need to construct some proxy for it from more accessible variables. There will be room for argument about how best to do this but agreement on the existence of the latent variable. In most social applications the latent variables do not have this status. Business sentiment is not something which exists in the sense that personal wealth does. It is a summarizing concept which comes after rather than before the indicators of it which we measure. Much of the philosophical debate which takes place on latent variable models centres on *reification*; that is, on speaking as though such things as quality of life and business sentiment were real entities in the sense that length and weight are. However, the usefulness and validity of the methods to be described in this book does not depend primarily on whether one adopts a realist or an instrumentalist view of latent variables. Whether one regards the latent variables as existing in some real world or merely as a means of thinking economically about complex relationships, it is possible to use the methods for prediction or establishing relationships *as if* the theory were dealing with real entities. In fact, as we shall see, some methods which appear to be purely empirical lead their users to behave as if they had adopted a latent variable model.

This book aims to unify a diverse body of statistical methods which have as their common feature a dependence on the notion of a latent variable. These include factor analysis, latent structure and latent trait analysis and the analysis of mixtures of distributions. We shall not treat them under their traditional headings since these tend to obscure their essential unity. Instead we shall approach the problem in a more general way allowing the various existing techniques (and some new ones) to emerge as special cases. We do this informally in the following section and then, more systematically, in Chapter 4.

1.2 A theoretical framework

There are two sorts of variables to be considered and they will be distinguished as follows. Variables which can be directly observed, also known as *manifest* variables, will be denoted by x. A collection of p manifest variables will be distinguished by subscripts and written as a column vector $\mathbf{x} = (x_1, x_2, \ldots, x_p)'$. In the interests of notational economy we shall not distinguish between random variables and the values which they take. In practice the distinction will be effected by the addition of a second subscript, thus x_{ih} will be the observed value of random variable x_i for the hth sample member and $\mathbf{x}_h$ will be that member's x-vector. Latent variables will be denoted by y and q will be the number of such variables. In practice q will be much smaller than p.

It is common to classify the level of measurement of variables as nominal, ordinal, interval or ratio. For our purposes we shall adopt a twofold classification: *metrical* and *categorical.* Metrical variables have realized values in the set of real numbers and may be discrete or continuous. Categorical variables assign individuals to one of a set of categories. They may be un-ordered or ordered; ordering commonly arises when the categories have been formed by grouping metrical variables. A useful way of seeing how the techniques mentioned above are related is to construct the fourfold classification of Table 1.1.

Table 1.1 Classification of latent variable methods

		Manifest variables	
		Metrical	Categorical
Latent variables	Metrical	factor analysis	latent trait analysis factor analysis of categorical data
	Categorical	latent profile analysis	latent class analysis
		analysis of mixtures	

Since both manifest and latent variables, by definition, vary from one individual to another they are represented in the theory by random variables. The relationships between them must therefore be expressed in terms of probability distributions, so that after the x's have been observed the information we have about $\mathbf{y}$ is given by its conditional distribution given $\mathbf{x}$. We shall set out the essentials as if both $\mathbf{x}$ and $\mathbf{y}$ were continuous, but the modifications for the other combinations of Table 1.1 are straightforward and do not bear upon the main point to be made.

As only $\mathbf{x}$ can be observed, any inference must be based on the joint distribution whose density may be expressed as

$$f(\mathbf{x}) = \int_{R_y} h(\mathbf{y})g(\mathbf{x} \mid \mathbf{y})\, d\mathbf{y} \tag{1.1}$$

where R_y is the range space of $\mathbf{y}$ (the subscript will usually be omitted). Our main interest is in what can be known about $\mathbf{y}$ after $\mathbf{x}$ has been observed. This information is conveyed by the conditional density

$$h(\mathbf{y} \mid \mathbf{x}) = h(\mathbf{y})g(\mathbf{x} \mid \mathbf{y})/f(\mathbf{x}). \tag{1.2}$$

The nature of the problem we face is now clear. In order to find $h(\mathbf{y} \mid \mathbf{x})$ we need to know both h and g, but all that we can estimate is f. It is obvious that h and g are not uniquely determined by (1.1) and thus, at this level of generality, we cannot obtain a complete specification of $h(\mathbf{y} \mid \mathbf{x})$. For progress to be made we must place some further restriction on the classes of functions to be considered. In fact (1.1) and (1.2) do not specify a model, they are merely different ways of expressing the fact that $\mathbf{x}$ and $\mathbf{y}$ are random variables that are mutually dependent on one another. No other assumption is involved. However, rather more is implied in our discussion than we have yet brought out. If the x's are each related to one or more of the y's then there will be correlations among the x's. Thus if x_1 and x_2 both depend on y_1 we may expect the common influence of y_1 to induce a correlation between x_1 and x_2. Conversely if x_1 and x_2 were uncorrelated there would be no grounds for supposing that they had anything in common. Taking this one step further, if x_1 and x_2 are uncorrelated *when y_1 is held fixed* we may infer that no other y is needed to account for their relationship since the existence of such a y would induce a correlation even if y_1 were fixed.

In general we are saying that if the correlations among the x's are induced by a set of latent variables $\mathbf{y}$, then when all y's are accounted for the x's will be uncorrelated if all the y's are held fixed. If this were not so the set of y's would not be *complete* and we should have to add at least one more. Thus q must be chosen so that

$$g(\mathbf{x} \mid \mathbf{y}) = \prod_{i=1}^{p} g_i(x_i \mid \mathbf{y}). \tag{1.3}$$

This is often spoken of as the assumption (or axiom) of conditional (or local) independence. But it is misleading to think of it as an assumption of the kind that could be tested empirically because there is no way in which $\mathbf{y}$ can be fixed and therefore no way in which the independence can be tested. It is better regarded as a definition of what we mean when we say that the set of latent variables $\mathbf{y}$ is complete. In other words, that $\mathbf{y}$ is sufficient to explain the dependencies among the x's. We are asking whether $f(\mathbf{x})$ admits the representation

$$f(\mathbf{x}) = \int_R h(\mathbf{y}) \prod_{i=1}^{p} g_i(x_i \mid \mathbf{y})\, d\mathbf{y} \tag{1.4}$$

for some q, h and $\{g_i\}$. In practice we are interested in whether (1.4) is an adequate representation for some small value of q. The dependence of (1.4) on q is concealed by the notation and is thus easily overlooked. We do not *assume* that (1.3) holds; our whole analysis is directed to discovering the smallest q for which such a representation is adequate.

In order to make much progress in this quest some further constraints must be imposed, but the point just made is so fundamental that we shall illustrate its meaning by taking a very simple and familiar example in which all variables are not continuous but binary. Suppose that we are presented with a 2×2 contingency table as follows.

	A	$\bar{A}$	
B	350	200	550
$\bar{B}$	150	300	450
	500	500	1000

Leaving aside questions of statistical significance, this table exhibits an association between the two variables. If A was being a heavy smoker and B was having lung cancer someone might object that the association was spurious and that it was attributable to some third factor with which A and B were both associated—such as living in an urban environment. If we go on to look at the association between A and B in the presence and absence of C we might obtain:

C	A	$\bar{A}$	
B	320	80	400
$\bar{B}$	80	20	100
	400	100	500

$\bar{C}$	A	$\bar{A}$	
B	30	120	150
$\bar{B}$	70	280	350
	100	400	500

The original association has now vanished and we conclude that the underlying variable C was wholly responsible for it.

Even in the absence of a suggestion about C it would still be pertinent to ask whether the original table could be decomposed into two tables exhibiting independence. If so we might then look at the members of each subgroup to see if they had anything in common, such as most of one group living in an urban environment. The idea can be extended to a p-way table and again we can enquire whether it can be decomposed into sub-tables in which the variables are independent. If this were possible there would be grounds for supposing that there was some latent categorization which fully explained the original association. The discovery of such a decomposition would amount to having found a latent categorical variable for which conditional independence held. The validity of the search does not require the assumption that the goal will be reached.

We now illustrate these rather abstract ideas by showing how they relate to two of the best-known latent variable models.

1.3 Binary manifest variables and a single binary latent variable

We begin with a development of the idea behind the numerical example given above by introducing a simple latent class model. Suppose there are p binary variables $x_1, x_2, \ldots, x_p$ with $x_i = 0$ or 1 for all i. Let us consider whether their mutual association could be accounted for by a single binary variable y. In other words, is it possible to divide the population into two parts so that the x's are mutually independent in each group? It is convenient to label the two hypothetical groups 1 and 0 (any other labelling would serve equally well). The prior distribution of y, h of the general theory, may then be written

$$h(1) = Pr\{y = 1\} = \eta \quad \text{and} \quad h(0) = 1 - h(1). \tag{1.5}$$

The conditional distribution of x_i given y, $g_i(x_i \mid y)$, will then be that of a Bernoulli random variable written

$$g_i(x_i \mid y) = Pr\{x_i \mid y\} = \pi_{iy}^{x_i}(1 - \pi_{iy})^{1-x_i}, \quad (x_i, y = 0, 1). \tag{1.6}$$

Notice that in this simple case the form of the distributions h and $\{g_i\}$ is not in question, it is only their parameters, η, $\{\pi_{i0}\}$ and $\{\pi_{i1}\}$ which are unspecified by the model.

For this model

$$f(\mathbf{x}) = \eta \prod_{i=1}^{p} \pi_{i1}^{x_i}(1 - \pi_{i1})^{i-x_i} + (1 - \eta) \prod_{i=1}^{p} \pi_{i0}^{x_i}(1 - \pi_{i0})^{1-x_i} \tag{1.7}$$

the integral of (1.1) becoming a sum. To test whether such a decomposition is adequate we would fit the probability distribution of (1.7) to the observed frequency distribution of **x**-vectors and apply a goodness of fit test. As we shall see later the parameters of (1.7) can be estimated by maximum likelihood. If the fit were not adequate we might go on to consider three or more latent classes or, perhaps, to allow y to be metrical. If the fit were satisfactory we might wish to have a rule for allocating individuals to one or other of the latent classes on the basis of their **x**-vector. For this we need the posterior distribution

$$h(1 \mid \mathbf{x}) = Pr\{y = 1 \mid \mathbf{x}\} = \eta \prod_{i=1}^{p} \pi_{i1}^{x_i}(1 - \pi_{i1})^{1-x_i} \Big/$$

$$\left\{\eta \prod_{i=1}^{p} \pi_{i1}^{x_i}(1 - \pi_{i1})^{1-x_i} + (1 - \eta) \prod_{i=1}^{p} \pi_{i0}^{x_i}(1 - \pi_{i0})^{1-x_i}\right\}$$

$$= 1 \Big/ \left[1 + \left(\frac{1-\eta}{\eta}\right) \exp \sum_{i=1}^{p} \left\{x_i \log \frac{\pi_{i0}}{\pi_{i1}} + (1 - x_i) \log \frac{1 - \pi_{i0}}{1 - \pi_{i1}}\right\}\right]. \tag{1.8}$$

Clearly individuals cannot be allocated with certainty, but if estimates of the parameters are available an allocation can be made on the basis of which group is more probable. Thus we would allocate to group 1 if

$$h(1 \mid \mathbf{x}) > h(0 \mid \mathbf{x}),$$

that is, if

$$X = \sum_{i=1}^{p} x_i \{\text{logit}\, \pi_{i0} - \text{logit}\, \pi_{i1}\}$$

$$> \sum_{i=1}^{p} \log\{(1 - \pi_{i1})/(1 - \pi_{i0})\} - \text{logit}\, \eta \qquad (1.9)$$

where logit $u = \log\{u/(1-u)\}$. An interesting feature of this result is that the rule for discrimination depends on the x's in a linear fashion. Here this is a direct consequence of the fact that the posterior distribution of (1.8) depends on $\mathbf{x}$ only through this same linear combination. In that sense X contains all the relevant information in the data about the latent variable. This is a key idea which is at the heart of the theoretical treatment of Chapter 4.

It is worth emphasizing again that much of the arbitrariness in the general approach with which we started has been avoided by fixing the number of latent classes and hence the form of the distribution h. There might, of course, be some prior grounds for expecting two latent groups, but nothing is lost by the assumption because if it fails we can go on to try more.

1.4 A model based on normal distributions

If $\mathbf{x}$ consists of metrical variables the writing down of a model is less straightforward. As before we might postulate two latent classes and then we should have

$$f(\mathbf{x}) = \eta \prod_{i=1}^{p} g_i(x_i \mid y = 1) + (1 - \eta) \prod_{i=1}^{p} g_i(x_i \mid y = 0) \qquad (1.10)$$

but then we are faced with the choice of conditional distributions for x_i. There is now no natural choice as there was when the x's were binary. We could, of course, make a plausible guess, fit the resulting model and try to justify our choice retrospectively by a goodness of fit test. Thus if a normal conditional distribution seemed reasonable we could proceed along the same lines as in the last section. Models constructed in this way will be discussed in Chapter 2.

Another approach is to start at the other end with $f(\mathbf{x})$ since this can be estimated from the data. Suppose our sample of $\mathbf{x}$'s could be regarded as coming from a multivariate normal distribution wth mean vector $\boldsymbol{\mu}$ and dispersion matrix $\boldsymbol{\Sigma}$. We might then ask whether the multivariate normal distribution admits a representation of the form (1.4) and if so whether it is unique. It is easy to find one such representation using standard results of distribution theory. Suppose that

$$\mathbf{y} \frown N_q(\mathbf{0}, \mathbf{I})$$

and

$$\mathbf{x} \mid \mathbf{y} \frown N_p(\boldsymbol{\mu} + \boldsymbol{\Lambda}\mathbf{y}, \boldsymbol{\psi}) \tag{1.11}$$

where $\boldsymbol{\Lambda}$ is a $p \times q$ matrix of coefficients and $\boldsymbol{\psi}$ is a diagonal matrix of variances. It then follows that

$$\mathbf{x} \frown N_p(\boldsymbol{\mu}, \boldsymbol{\Lambda}\boldsymbol{\Lambda}' + \boldsymbol{\psi}) \tag{1.12}$$

which is of the required form. Note that although this representation is possible for all $q \leqslant p$ it does not imply that a $\boldsymbol{\Lambda}$ and $\boldsymbol{\psi}$ can be found such that $\boldsymbol{\Lambda}\boldsymbol{\Lambda}' + \boldsymbol{\psi}$ is equal to the given $\boldsymbol{\Sigma}$ unless $q = p$. Every model specified by (1.11) leads to a multivariate normal $\mathbf{x}$, but if $q < p$ the converse is not true. The point of the argument is to show that the model of (1.11) is worth entertaining if the x's have a multivariate normal distribution.

The posterior distribution of $\mathbf{y}$ is easily obtained by standard methods and it, too, turns out to be normal. Thus,

$$\mathbf{y} \mid \mathbf{x} \frown N_q\{\boldsymbol{\Lambda}'\boldsymbol{\Sigma}^{-1}(\mathbf{x} - \boldsymbol{\mu}), \quad (\boldsymbol{\Lambda}'\boldsymbol{\psi}^{-1}\boldsymbol{\Lambda} + \mathbf{I})^{-1}\} \tag{1.13}$$

where $\boldsymbol{\Sigma} = \boldsymbol{\Lambda}\boldsymbol{\Lambda}' + \boldsymbol{\psi}$ and $q < p$. The mean of this distribution might then be used to predict $\mathbf{y}$ for a given $\mathbf{x}$ and the precision of the predictions would be given by the elements of the dispersion matrix.

Unfortunately the decomposition of (1.11) is not unique as we now show. The point is a general one which applies whether or not $\mathbf{x}$ is normal.

Suppose that $\mathbf{y}$ is continuous and that we make a 1–1 transformation of the factor space from $\mathbf{y}$ to $\mathbf{v}$. This will have no effect on $f(\mathbf{x})$ since it is merely a change of variable in the integral of (1.1), but both of the functions h and g will be changed. In the case of h the *form* of the prior distribution will, in general, be different, and in the case of g there will be a change, for example, in the regression of $\mathbf{x}$ on the latent variables. It is thus clear that there is no unique way of expressing $f(\mathbf{x})$ as in (1.4) and therefore no empirical means of distinguishing among the possibilities. We are thus not entitled to draw conclusions from any analysis which would be vitiated by a transformation of the latent space. However, there may be some representations which are easier to interpret than others. We note in the present case, from (1.11), that the regression of x_i on $\mathbf{y}$ is linear and this enables us to interpret the elements of $\mathbf{\Lambda}$ as weights determining the effect of each $\mathbf{y}$ on a particular x_i. Any non-linear transformation of $\mathbf{y}$ would destroy this relationship.

Another way of looking at the matter is to argue that the indeterminacy of h leaves us free to adopt a metric for $\mathbf{y}$ such that h has some convenient form. A normal scale is familiar so we might require each y_i to have a standard normal distribution. If, as a further convenience, we make the y's independent as in (1.11) then the form of g_i is uniquely determined and we would then note that we had the additional benefit of linear regressions.

In general, if we find the representation (1.4) is possible, we may fix either h or $\{g_i\}$; in the normal case either approach leads us to (1.11).

If $\mathbf{x}$ is normal there is an important transformation which leaves the form of h unchanged and which thus still leaves a degree of arbitrariness about $\{g_i\}$. Suppose $q \geqslant 2$ then the orthogonal transformation $\mathbf{v} = \mathbf{M}\mathbf{y}$, $(\mathbf{M}'\mathbf{M} = \mathbf{I})$ gives

$$\mathbf{v} \frown N_q(\mathbf{0}, \mathbf{I})$$

which is the same distribution as $\mathbf{y}$ had. The conditional distribution is now

$$\mathbf{x} \mid \mathbf{v} \frown N_p(\boldsymbol{\mu} + \mathbf{\Lambda}\mathbf{M}'\mathbf{v}, \boldsymbol{\psi}) \tag{1.14}$$

so that a model with weights $\mathbf{\Lambda}$ is indistinguishable from one with weights $\mathbf{\Lambda}\mathbf{M}'$. The effect of orthogonally transforming the latent space is thus exactly the same as transforming the weight matrix. The joint distribution of $\mathbf{x}$ is, of course, unaffected by this. In the one case the dispersion matrix is $\mathbf{\Lambda}\mathbf{\Lambda}' + \boldsymbol{\psi}$ and in the other it is $\mathbf{\Lambda}\mathbf{M}'\mathbf{M}\mathbf{\Lambda}' + \boldsymbol{\psi}$ and these are equal because $\mathbf{M}'\mathbf{M} = \mathbf{I}$.

The indeterminacy of the factor model revealed by this analysis has advantages and disadvantages. So far as determining q, the dimensionality of the latent space, is concerned, there is no problem. But from a purely statistical point of view the arbitrariness is unsatisfactory and in the next chapter we shall consider how it might be removed. However there may be practical advantages in allowing the analyst freedom to choose from among a set of transformations that which has most substantive meaning. This too is a matter to which we shall return.

The reader already familiar with factor analysis will have recognized many of the formulae in this section, even though the setting may be unfamiliar. The usual treatment, to which we shall come later, starts with the linear regression implicit in (1.11) and then adds the distributional assumptions in a convenient but more or less arbitrary way. In particular, the essential role of the conditional independence postulate is thereby obscured. The advantage of starting with the distribution of $\mathbf{x}$ is that it leads to the usual model in a more compelling way but, at the same time, makes the essential arbitrariness of some of the usual assumptions clearer. We shall return to these points in Chapter 4 where we shall see that the present approach lends itself more readily to generalization when the rather special properties of normal distributions which make the usual linear model the natural one are no longer available.

Principal Components

We remarked above that the representation

$$\mathbf{\Sigma} = \mathbf{\Lambda\Lambda}' + \boldsymbol{\psi} \tag{1.15}$$

is always possible when $q = p$. This follows from the fact that $\mathbf{\Sigma}$ is a symmetric matrix and so can be expressed as

$$\mathbf{\Sigma} = \mathbf{A\Theta A}'$$

where $\mathbf{\Theta}$ is a diagonal matrix whose elements are the eigenvalues of $\mathbf{\Sigma}$ and $\mathbf{A}$ is an orthogonal matrix whose columns are the corresponding eigenvectors of $\mathbf{\Sigma}$. Consequently if we choose

$$\mathbf{\Lambda} = \mathbf{A\Theta}^{\frac{1}{2}} \quad \text{and} \quad \boldsymbol{\psi} = \mathbf{0}$$

(1.15) follows. The conditional distribution of $\mathbf{x}$ given $\mathbf{y}$ in (1.11) is now degenerate with all the probability concentrated at the mean given by

$$\mathbf{x} = \boldsymbol{\mu} + \mathbf{\Lambda y} = \boldsymbol{\mu} + \mathbf{A\Theta}^{\frac{1}{2}}\mathbf{y} \tag{1.16}$$

with $\mathbf{x}$ having been expressed exactly as a linear combination of independent standard normal variables. The variables

$$\mathbf{y}^* = \mathbf{\Theta}^{\frac{1}{2}}\mathbf{y} = \mathbf{A}'(\mathbf{x} - \boldsymbol{\mu}) \tag{1.17}$$

are known as the *principal components*. If, say, q of the eigenvalues of $\mathbf{\Sigma}$ are large and the remainder small then those components associated with the dominant eigenvalues will be mainly responsible for the correlation structure of the x's, and the effect of the remainder will be like that of the error term in a factor model with q factors. The fact that a principal component analysis may yield results similar to a factor analysis can be turned to advantage as we shall see later.

In the usual derivation of principal components there is no need to make distributional assumptions because (1.16) can be viewed as a relationship between mathematical variables. There are several ways of motivating the theory. One is to choose y_1^* so that it accounts for as large a proportion of the total variance of $\mathbf{x}$ (measured by $\Sigma\, \text{var}(x_i)$) as possible; y_2^* is then required to account for as much of the remaining variance as possible, subject to being uncorrelated with y_1^* and so on with each successive component being uncorrelated with its predecessors and accounting for as much residual variance as possible. This leads to (1.17) with $\text{var}(y_i^*) = \theta_i$ $(i = 1, 2, \ldots, p)$.

1.5 Historical note

The origins of latent variable modelling can be traced back to the early part of the present century, notably in the study of human abilities by Burt and Spearman. They were concerned with the fact that people, especially children, who performed well in one test of mental ability also tended to do well in others. This led to the idea that all an individual's scores were manifestations of some underlying general ability which might be called general intelligence or "g". However, the scores on different items were certainly not perfectly correlated and this was explained by invoking factors specific to each item to account for the variation in performance from one item to another. The general and specific factors were thus supposed to combine in some way to produce the actual performance. The simplest way in which this might happen would be for the two effects to be independent and additive. Thus one could postulate a model of the form

$$x_i = \mu_i + \lambda_i y + e_i \tag{1.18}$$

where x_i is a random variable representing the score on test i, y the general factor and e_i the factor specific to item i. (The notation has been chosen to conform to that used in the early part of this chapter.) If y and the e's are mutually independent the correlation coefficient of x_i and x_j will have the form

$$\rho_{ij} = a_i a_j \tag{1.19}$$

and a rough test of the plausibility of the model can be made by inspection of the correlation matrix. Thus it follows from (1.19) that

$$\rho_{ih}/\rho_{ij} = a_h/a_j. \tag{1.20}$$

If we form the ratios on the left-hand side for any pair of columns of the matrix they should be constant across rows. Another way of exhibiting the pattern in the matrix is by observing that if $l_{ij} \neq 0$

$$\rho_{ih}\rho_{hj}/\rho_{ij} = a_h^2 \tag{1.21}$$

for all pairs i and j. Since

$$a_h^2 = \lambda_h^2\, \text{var}(y)/\{\lambda_h^2\, \text{var}(y) + \text{var}(e_h)\}$$

it is thus possible to estimate the ratio $\lambda_h^2 \operatorname{var}(y)/\operatorname{var}(e_h)$ and so determine the relative contributions of the general and specific factors to the observed score.

If the correlation matrix fails to exhibit the simple pattern required by (1.19) it is natural to add further general factors to the model in the hope of reproducing the observed pattern more exactly.

Within psychology there was a new impetus in the 1930's from Thurstone and his associates in Chicago. He advocated the desirability of obtaining solutions possessing what he called "simple structure" which meant replacing Spearman's single general factor by a cluster of distinct but correlated factors representing different abilities. The general factor of earlier work was thus seen as an "average" of the "real" factors and thus of little substantive meaning. Much of the work in this tradition was published in the then new journal *Psychometrika,* the fiftieth anniversary of which was the occasion of a historical review by Mulaik (1986). Although this was primarily concerned with the subject as it has developed in the pages of that journal there are indications of what was happening on a broader front. The article stops short of the most recent advances some of which are in line with Mulaik's prognostications.

Thurstone's writings are now mainly of historical interest, but his book *Multiple Factor Analysis* (Thurstone, 1947) is still worth reading for its insight into the essential character and purpose of factor analysis—something which easily becomes overlaid by technical details. The same is true of Thomson's (1939) *The Factorial Analysis of Human Ability* which anticipates many topics which have occupied post-war theorists.

Two more recent books in the psychological tradition are those by Mulaik (1972) and Harman (1976). Mulaik gives a comprehensive statement of the theory as it then existed, interspersed with illuminating historical material. It is still a useful source of many of the basic algebraic derivations. The book by Harman, now in its third edition, lays considerable emphasis on statistical methods and computational matters but with less mathematical apparatus than Mulaik.

A major treatment of an entirely different kind will be found in Cattell (1978) who is at pains to emphasize the limitations of a purely statistical approach to factor analysis as against its use as part of scientific method. His exposition is made without any explicit use of models.

The present book lies at the opposite pole to Cattell in that it gives priority to modelling. A model serves to clarify the conceptual basis of the subject and provides a framework for analysis and interpretation. In a subject which has been criticized as arbitrary and hence too subjective, it is especially necessary to clarify in as rigorous a way as possible what is being assumed and what can be legitimately inferred. The general approach used here goes back to Anderson (1955) but his work pre-dated the computer era and the time was not ripe for its exploitation.

The major treatment of factor analysis within the statistical tradition is due to Lawley and Maxwell (1971) whose book remains unrivalled as a source of results on the normal theory factor model—especially in the area of estimation and hypothesis testing. More recent work associated with Bentler, Browne, Jöreskog, McDonald and others, may be regarded as a generalization of the factor model and goes under the name of the analysis of covariance structures. This is based on what are called linear structural equation models which incorporate not only the basic factor model but also linear relationships among the y's (or factors). Our general framework can be extended to include such models, but this is outside the scope of the present book.

Statisticians have often preferred Principal Components Analysis to Factor Analysis. The idea of this technique goes back to Karl Pearson (1901) but it was developed as a multivariate technique by Hotelling (1933). Although it is quite distinct from Factor Analysis in that it does not depend on any probability model, the close affinities between the two techniques outlined above mean that they are often confused.

Latent structure analysis arises from models in which the latent variables are categorical. Its development has, until recently, been entirely separate from factor analysis. It originated with Lazarsfeld as a tool for sociological analysis and was expounded in the book by Lazarsfeld and Henry (1968). Here again modern computing facilities have greatly extended the applicability of the methods. A recent, more up-to-date account, is in Everitt (1984).

A third, distinct, strand which has contributed to the current range of models comes from educational testing. The manifest variables in this case are usually categorical, often indicating whether an individual got an item right or wrong in a test. There is assumed to be a single continuous latent variable corresponding to an ability of some kind, and the practical objective is to locate individuals on some suitable scale. The contribution of Birnbaum to Lord and Novick (1968) was a major step forward in this field which is now the scene of substantial research activity and some controversy. Much of the latter has centred upon the so-called Rasch model (see, for example, Rasch (1960) or Andersen (1980)) which has many appealing statistical properties. However, a feature of much of this work on latent trait models, as they are called, which tends to obscure their connection with the general family set out in Section 1.2, is that the latent traits are not treated as random variables. Instead a parameter is introduced to represent each individual's position on the latent scale. Such a model may be termed a "fixed effects" model by analogy with the usage in other branches of statistics. Its use would be appropriate if we were interested only in the abilities of the particular individuals in the sample and not in the distribution of ability in any population from which they might have been drawn. It is not easy to find practical examples where this would be the case, and for this reason such models will receive only passing mention in the rest of the book. Fixed effects models have also been considered in factor analysis by Whittle (1953) and Anderson and Rubin (1956), for example, but have not attracted much practical interest. One of their chief disadvantages is that the number of parameters goes up in proportion to the sample size and this creates problems with the behaviour of maximum likelihood estimators. However, there are circumstances in which such methods are relatively simple and can be made to yield estimates of the item parameters which are virtually the same as those derived from a random effects model. They thus have a certain practical interest but in spite of a voluminous and often polemical literature they are, from our standpoint, outside the mainstream of theoretical development.

The time is ripe for a reappraisal of this area of statistics. The combination of a unified theoretical treatment of the whole field with the flexibility and power of current computing facilities offers ample scope for new developments in many branches of social and natural science. There are still many gaps in the theory, especially on the inferential side, and much that is done in practice (even by some of the major software packages) is without adequate theoretical justification. But this is an argument for better theory and not one for abandoning what we have.

CHAPTER 2

Latent Class Models

2.1 Introduction

In Chapter 1 we gave examples of two types of latent variable model. One, discussed in Section 1.4, provided a first look at the traditional linear model of factor analysis and this is the main subject of the next chapter. The second was a simple example of a latent class model which forms the starting-point of the present chapter. We shall extend that model in three principal directions as follows:

(a) by allowing more than two latent classes
(b) by allowing the manifest variables to be polytomous
(c) by allowing the manifest variables to be metrical.

The first systematic treatment of latent class models was by Lazarsfeld and Henry (1968). They tended to emphasize the differences between factor analysis and latent structure analysis remarking (p. 6) that "Latent structure analysis is in some ways similar and in many very basic ways different from factor analysis". This judgement, which perhaps had greater justification 20 years ago than today, may have inhibited the development of a common approach to all latent variable models which this book sets out to achieve. The introduction of efficient estimation techniques using modern computers and a broader statistical perspective have combined to render much of that book obsolete, but it remains a path-breaking effort in the statistical treatment of social data. The more recent work of Goodman (1978, Part Four) contains many interesting developments, though the connections with other latent variable methods are obscured by choice of notation. Everitt's (1984) book on latent structure models provides a good elementary introduction to the topic.

As a preparatory step we shall describe briefly three studies in which latent class models have been used. Further examples will be introduced later as illustrations of the methods. The first is taken from the study by Aitkin, Anderson and Hinde (1981) of an enquiry into teaching styles which was reported in Bennett (1976). This study has been much debated and the data have been re-analysed several times. The use of a latent class model arose out of the need to see whether teachers could be classified according to their style of teaching—for example, traditional or progressive. The manifest variables consisted of 38 binary variables arising from questions relating to such things as whether pupils had a choice of where to sit, whether they were allowed to talk freely, whether pupils' work was marked or graded, and whether there were many children who create discipline problems. One might then have wished to use the model of Section 1.3 to see whether each of the 468 teachers could be allocated to one of two classes on the basis of their replies to the 38 questions. If this were not possible three or more latent classes might be tried.

The second example arises in educational testing. What are called criterion referenced tests are sometimes constructed to test whether children have mastered particular skills or concepts. Such a test will consist of a set of items each designed to test some aspect of mastery of the skill in question. One then has to decide whether a child is a "master" or a "non-master" on the basis of which of the items they got right or wrong. The simplest possible model would be one according to which masters got all items right and non-masters got them all wrong. In practice such clear-cut distinctions are not possible and so Macready and Dayton (1977) and others have introduced a latent class model in which non-masters sometimes get items right by guessing and masters sometimes get them wrong by carelessness or forgetfulness. If guessing and forgetting have fixed probabilities, different for each item, we have a model of exactly the form discussed in Section 1.3. If it fits the data satisfactorily it could be used for allocating pupils to mastery classes.

The third example is a study of nearly 50,000 Scottish births carried out by Pickering and Forbes (1984). Eleven categorical variables were recorded for each baby; eight were binary concerning, for example, whether the baby had convulsions or jaundice and three, including birth weight, had more than two categories. Here the aim was to identify classes of abnormal birth since if these showed different geographical incidence, resources needed to cope with them could be distributed more efficiently. The expectation was that there would be one (large) "normal" class together with an unknown number of abnormal classes.

2.2 Latent class models with binary manifest variables

The two-class model may easily be extended to K classes as set out below. In some cases K may be specified by the theory being tested, as in the case of the mastery model, but usually it will be an unknown to be determined.

Let π_{ij} be the probability of a positive response on variable i for a person in category j ($i = 1, 2, \ldots, p; j = 0, 1, \ldots, c_i - 1$) and let η_j be the prior probability that a randomly chosen individual is in class j $\left(\sum_{j=0}^{K-1} \eta_j = 1\right)$. For the case of K classes (1.7) becomes

$$f(\mathbf{x}) = \sum_{j=0}^{K-1} \eta_j \prod_{i=1}^{p} \pi_{ij}^{x_i}(1 - \pi_{ij})^{1-x_i}. \tag{2.1}$$

The posterior probability that an individual with response vector $\mathbf{x}$ belongs to category j is thus

$$h(j \mid \mathbf{x}) = \eta_j \prod_{i=1}^{p} \pi_{ij}^{x_i}(1 - \pi_{ij})^{1-x_i}/f(\mathbf{x}) \quad (j = 0, 1, \ldots, K-1). \tag{2.2}$$

We can use (2.2) to construct an allocation rule according to which an individual is placed in the class for which the posterior probability is greatest. The principal statistical problem is thus the estimation of the parameters and testing goodness of fit. On the substantive side the main problem is to identify the latent classes—that is to interpret them in terms which make practical sense. In this connection it is worth noting that our model includes the case where there are two or more cross-classified latent categorical variables. For example, if there were two binary cross-classified latent categorical variables, their joint distribution would involve four classes representing the possible combinations of two factors at two levels. If, in such a case, we wished to require the latent variables to be independent this would impose a constraint on the η's.

Maximum Likelihood Estimation

The log-likelihood function derived from (2.1) is complicated, but it can be maximized using standard optimization routines. McHugh (1956 and 1958) showed how this might be done using the standard Newton–Raphson technique. However, as with many other latent variable models, an easier method which enables larger problems to be tackled is offered by the E-M algorithm. The basic reference is Dempster, Laird and Rubin (1977) supplemented by Wu (1983), but a version suitable for the latent class model was given by Goodman (1978). Here we shall develop the method from first principles in a manner which gives some insight into the nature of the estimators.

From (2.1) we find the log-likelihood for a random sample of size n to be

$$L = \sum_{h=1}^{n} \log\left\{\sum_{j=0}^{K-1} \eta_j \prod_{i=1}^{p} \pi_{ij}^{x_{ih}}(1 - \pi_{ij})^{1-x_{ih}}\right\}. \tag{2.3}$$

This has to be maximized subject to $\sum \eta_j = 1$, so we find the unrestrained maximum of

$$\phi = L + \theta \sum_{j=0}^{K-1} \eta_j$$

where θ is an undetermined multiplier. Finding partial derivatives, we have

$$\frac{\partial \phi}{\partial \eta_j} = \sum_{h=1}^{n} \left\{ \prod_{i=1}^{p} \pi_{ij}^{x_{ih}} (1-\pi_{ij})^{1-x_{ih}} / f(\mathbf{x}_h) \right\} + \theta \quad (j = 0, 1, \ldots, K-1)$$

$$= \sum_{h=1}^{n} \{ g(\mathbf{x}_h \mid j) / f(\mathbf{x}_h) \} + \theta \qquad (2.4)$$

where $g(\mathbf{x}_h \mid j)$ is the joint probability of $\mathbf{x}_h$ for an individual in class j;

$$\frac{\partial \phi}{\partial \pi_{ij}} = \sum_{h=1}^{n} \eta_j \frac{\partial}{\partial \pi_{ij}} g(\mathbf{x}_h \mid j) / f(\mathbf{x}_h) \quad (i = 1, 2, \ldots, p; j = 0, 1, \ldots, K-1).$$

Now

$$\frac{\partial g(\mathbf{x}_h \mid j)}{\partial \pi_{ij}} = \frac{\partial}{\partial \pi_{ij}} \exp \sum_{i=1}^{p} \{ x_{ih} \log \pi_{ij} + (1 - x_{ih}) \log(1 - \pi_{ij}) \}$$

$$= g(\mathbf{x}_h \mid j) \left\{ \frac{x_{ih}}{\pi_{ij}} - \frac{(1 - x_{ih})}{(1 - \pi_{ij})} \right\}$$

$$= (x_{ih} - \pi_{ij}) g(\mathbf{x}_h \mid j) / \pi_{ij}(1 - \pi_{ij}).$$

Therefore,

$$\frac{\partial \phi}{\partial \pi_{ij}} = \{ \eta_j / \pi_{ij}(1 - \pi_{ij}) \} \sum_{h=1}^{n} (x_{ih} - \pi_{ij}) g(\mathbf{x}_h \mid j) / f(\mathbf{x}_h). \qquad (2.5)$$

The resulting equations can be simplified by expressing (2.4) and (2.5) in terms of the posterior probabilities $\{h(j \mid \mathbf{x})\}$. By Bayes' theorem,

$$h(j \mid \mathbf{x}) = \eta_j g(\mathbf{x}_h \mid j) / f(\mathbf{x}_h). \qquad (2.6)$$

Substituting in (2.4) and setting equal to zero, we find

$$\sum_{h=1}^{n} h(j \mid \mathbf{x}_n) = \theta \eta_j.$$

Summing both sides over j and using $\sum \eta_j = 1$ gives $\theta = n$, and hence the first estimating equation is

$$\hat{\eta}_j = \sum_{h=1}^{n} h(j \mid \mathbf{x}_h) / n \quad (j = 0, 1, \ldots, K-1). \qquad (2.7)$$

The second is

$$\sum_{h=1}^{n} (x_{ih} - \pi_{ij})h(j \mid \mathbf{x}_h)/\pi_{ij}(1 - \pi_{ij}) = 0$$

whence

$$\hat{\pi}_{ij} = \sum_{h=1}^{n} x_{ih}h(j \mid \mathbf{x}_h) \bigg/ \sum_{h=1}^{n} h(j \mid \mathbf{x}_h)$$

$$= \sum_{h=1}^{n} x_{ih}h(j \mid \mathbf{x}_h)/n\hat{\eta}_j \quad (i = 1, 2, \ldots, p; j = 0, 1, \ldots, K-1). \quad (2.8)$$

Although these equations have a simple form it must be remembered that $h(j \mid \mathbf{x}_h)$ is a complicated function of $\{\eta_j\}$ and $\{\pi_{ij}\}$ given by

$$h(j \mid \mathbf{x}_h) = \eta_j \prod_{i=1}^{p} \pi_{ij}^{x_{ih}}(1 - \pi_{ij})^{1-x_{ih}} \bigg/ \sum_{k=0}^{K-1} \eta_j \prod_{i=1}^{p} \pi_{ik}^{x_{ih}}(1 - \pi_{ik})^{1-x_{ih}}. \quad (2.9)$$

However, if $h(j \mid \mathbf{x}_h)$ were known it would be easy to solve (2.7) and (2.8) for $\{\eta_j\}$ and $\{\pi_{ij}\}$. The E-M algorithm takes advantage of this fact proceeding in a "zig-zag" fashion as follows:

(i) Choose an initial set of posterior probabilities $\{h(j \mid \mathbf{x}_h)\}$.
(ii) Use (2.7) and (2.8) to obtain a first approximation to $\{\hat{\eta}_j\}$ and $\{\hat{\pi}_{ij}\}$.
(iii) Substitute these estimates into (2.9) to obtain improved estimates of $\{h(j \mid \mathbf{x}_h)\}$.
(iv) Return to (ii) to obtain second approximations to the parameters and continue the cycle until convergence is attained.

The solution reached will be a local maximum (or saddle-point). It is known that models of this kind may have multiple maxima and the risk of this appears to increase as K, the number of classes, increases and to decrease with increasing sample size. Aitkin *et al.* (1981) provided an illustration of multiple maxima arising in the teaching styles data with only three latent classes. By using different starting values one can guard against the risk of mistaking a local for a global maximum, but if such multiple maxima do occur it is not clear what interpretation should be placed on the different sets of latent classes implied by the various local maxima.

A reasonable way of starting the iteration is to allocate individuals, arbitrarily, to latent classes on the basis of their total score $\left(\sum_{i=1}^{p} x_i\right)$. That is to take $h(j \mid \mathbf{x}_h) = 1$ if $\mathbf{x}_h$ is allocated to class j and $h(j \mid \mathbf{x}_h) = 0$ otherwise. Although the method may take a very large number of iterations to converge, the steps are simple and fast, so the total computing time need not be excessive. As well as providing parameter estimates the method also provides the posterior probabilities that each individual belongs to a given latent class. It does not provide the second derivatives needed for the calculation of standard errors but these can easily be found and evaluated at the solution point as we now see.

Standard Errors and Goodness of Fit

The second derivatives and cross-derivatives of L can be expressed in terms of the posterior distribution as follows

$$\frac{\partial^2 L}{\partial \eta_j \partial \eta_k} = -\sum_{h=1}^{n} h(j \mid \mathbf{x}_h) h(k \mid \mathbf{x}_h) / \eta_j \eta_k \tag{2.10}$$

$$\frac{\partial^2 L}{\partial \pi_{ij}\, \partial \pi_{lk}} = \sum_{h=1}^{n} (x_{ih} - \pi_{ij})(x_{lh} - \pi_{lk}) h(j \mid \mathbf{x}_h)\{\delta_{jk}(1 - \delta_{il}) - h(k \mid \mathbf{x}_h)\} / \pi_{ij}(1 - \pi_{ij})\pi_{lk}(1 - \pi_{lk}) \tag{2.11}$$

$$\frac{\partial^2 L}{\partial \eta_j\, \partial \pi_{ik}} = \sum_{h=1}^{n} (x_{ih} - \pi_{ij}) h(j \mid \mathbf{x}_h)\{\delta_{jk} - h(k \mid \mathbf{x}_h)\} \tag{2.12}$$

$(j, k = 0, 1, \ldots, K-1; i, l = 1, 2, \ldots, p)$.

The asymptotic variance–covariance matrix of the estimates is then the inverse of the expectation of the $K(p+1) \times K(p+1)$ matrix of the negatives of the derivatives set out above. The exact computation of the expected values involves summation over the 2^p possible score patterns $\mathbf{x}$. This is feasible if p is small, but with large p the number of terms becomes extremely large and the magnitude of each term so small that accurate calculation becomes impossible. In this case the expectation can be approximated by taking the inverse of the observed second derivative matrix. (We shall meet the same problem again in Chapter 6 where satisfactory checks on the adequacy of this approximation are reported.)

Other methods of investigating sampling error such as those based on the bootstrap or jackknife idea, which we shall discuss later, do not appear to have been used for latent class models, though some simulation results are reported in Everitt (1984). Few of the published applications of the latent class model report standard errors of their parameter estimates, though several attempt to judge the goodness of fit of the model. One way to do this is to compare the observed frequencies of the response patterns with those predicted by the model using a standard chi-squared or likelihood ratio goodness of fit test. This method was used by Goodman (1978) some of whose results are reported below. However, the method is only practicable if p is small. With n individuals and p variables the average number of cases per response pattern will be $n/2^p$, and even for very large samples this quickly reaches a level where many expected frequencies will be far too small to justify the distributional assumptions of the test. Even if cells with small expected frequencies are amalgamated, a point is soon reached where the number of degrees of freedom calculated in the conventional way becomes negative. With large values of p such as occurred in the teaching styles investigation there are about a million million cells only 468 of which are occupied, each by a single individual. A normal goodness of fit test is out of the question. Aitkin *et al.* (1981) use a graphical method based on the shape of the distribution of the total score (Σx_i), but the problem awaits a more satisfactory solution.

Allocation of Individuals to Latent Classes

In Section 1.3 we noted that the posterior probability distribution of class membership depended on a single linear function of the x's and hence that the allocation of individuals to classes could be based on this function. With more than two classes the relative posterior probability for classes j and k is

$$h(j \mid \mathbf{x}_h)/h(k \mid \mathbf{x}_h) = (\eta_j/\eta_k) \exp \sum_{i=1}^{p} [\{x_{ih} \log \pi_{ij} + (1 - x_{ih}) \log(1 - \pi_{ij})\} - \{x_{ih} \log \pi_{kj} + (1 - x_{ih}) \log(1 - \pi_{kj})\}].$$

If we allocate to the class with the larger posterior probability we thus do so on the basis of the value of

$$\sum_{i=1}^{p} \{x_{ih} \log \pi_{il} + (1 - x_{ih}) \log(1 - \pi_{il})\} - \log \eta_l \quad (l = j, k). \quad (2.13)$$

The most probable class in the complete set is the one for which (2.13) is a maximum when l ranges over the values $0, 1, \ldots, K-1$. The single discriminant function given in (1.9) for the binary case was, in effect, the difference between two expressions like those in (2.13).

Interpretation

Having established that a latent class model fits the data for some K, it may be necessary to interpret, that is name, the classes. In a case such as the criterion-referenced testing problem the classes are specified by the theory and it is only necessary to check that individuals allocated by the model to the class of "master" do have response patterns appropriate to that title. With less specific guidance from theory a judgement has to be made on the basis of the empirical evidence. Essentially we have to ask what those allocated to a given class have in common that distinguishes them from members of other classes. A way to do this is to look at the estimates of the probabilities $\{\pi_{ij}\}$. A useful pointer can be had, for particular j, by looking at those i for which $\hat{\pi}_{ij}$ is near 1 and those for which it is near zero. The former group will consist of attributes which members of class j are very likely to possess, the latter of those which they rarely have. The line of demarcation between the two groups will help to characterize the class, as we shall see in the following examples.

Panel Data on Student Attitudes

Our first example is based on Goodman's (1978) re-analysis of Coleman's panel data on student attitudes (Coleman (1964)). A total of 3398 boys were interviewed at two points in time about whether or not they considered themselves to be in "the leading crowd" (yes scored 1, no scored 0). At the same time they were asked whether they thought that such membership involved sometimes going against their principles (0 for yes and 1 for no). The frequency distribution across response patterns is given in Table 2.1 with fitted values obtained from two latent class models. The order in which responses are listed is: perceived membership at 1st interview, attitude at 1st

Table 2.1 The fit of two latent class models to Coleman's panel data

Response pattern	Observed frequency	Expected frequency with: 2 latent classes	4 latent classes
1111	458	408.3	454.8
1110	140	199.7	144.2
1101	110	94.3	109.1
1100	49	68.5	48.9
1011	171	227.7	172.3
1010	182	112.7	179.7
1001	56	77.9	58.3
1000	87	63.8	85.8
0111	184	159.7	188.6
0110	75	97.1	68.8
0101	531	403.0	530.5
0100	281	397.3	283.1
0011	85	110.6	82.1
0010	97	76.2	101.5
0001	338	451.5	337.3
0000	554	449.6	553.1

interview, perceived membership at 2nd interview, attitude at 2nd interview.

The first model supposes that there are two latent classes and the model parameters estimated by maximum likelihood are

$$\hat{\eta} = .40, \quad \hat{\pi}_{11} = .77, \quad \hat{\pi}_{21} = .64, \quad \hat{\pi}_{31} = .89, \quad \hat{\pi}_{41} = .67$$

$$\hat{\pi}_{10} = .10, \quad \hat{\pi}_{20} = .47, \quad \hat{\pi}_{30} = .09, \quad \hat{\pi}_{40} = .50$$

The log-likelihood statistic for testing the fit of this model whose expected frequencies are given in the third column of Table 2.1 has the value 249.50 which, on 6 degrees of freedom, is highly significant. There is therefore little point in trying to interpret the latent classes of this model.

The last column of the table shows the expected frequencies for a 4-class model, and here the fit is very much better with the likelihood ratio statistic being 1.27 on 4 degrees of freedom. However, this is a restricted model obtained by supposing that there are two binary latent classes giving four latent categories altogether. Goodman proposed the hypothesis that one of the binary latent variables would explain the association between the perceived membership at the two times and the second the association between the attitudes. Let us re-label the latent classes as follows

1	2	3	4
(0, 0)	(0, 1)	(1, 0)	(1, 1)

and denote the first latent binary variable by M (for membership) and the second by A (for attitude). According to the hypothesis the level of A does not affect variables 1 and 3 and M does not affect variables 2 and 4. This implies that

$$\pi_{i(1,0)} = \pi_{i(1,1)}; \quad \pi_{i(0,1)} = \pi_{i(0,0)} \quad \text{if } i = 1 \text{ or } 3$$

$$\pi_{i(0,0)} = \pi_{i(1,0)}; \quad \pi_{i(1,1)} = \pi_{i(0,1)} \quad \text{if } i = 2 \text{ or } 4.$$

This reduces the total number of parameters to 8 and this leaves $16 - 8 - 3 - 1 = 4$ degrees of freedom for testing goodness of fit. The maximum likelihood estimates of the η's and π's under the above constraints are

$$\hat{\eta}_1 = .27, \quad \hat{\eta}_2 = .13, \quad \hat{\eta}_3 = .23, \quad \hat{\pi}_4(=1 - \hat{\eta}_1 - \hat{\eta}_2 - \hat{\eta}_3) = .37$$

i	(0, 0)	(0, 1)	(1, 0)	(1, 1)
1	.75	.75	.11	.11
2	.81	.27	.81	.27
3	.91	.91	.08	.08
4	.83	.30	.83	.30

It is clear that a positive response to items 1 and 3 is common for people in latent classes (0, 0) and (0, 1) and rare if they are in classes (1, 0) and (1, 1). For items 2 and 4 it is those in classes (0, 0) and (1, 0) who have the high probability of response. There is thus little difficulty in identifying the latent variable M with a time-independent perceived membership of the leading crowd, which gives rise to a high positive association between the answers given to the same question at different times. Likewise A would refer to attitude to membership of this group.

The bivariate distribution of M and A is estimated to be

		M 0	1
A	0	.27	.23
	1	.13	.37

Yule's coefficient of association for this distribution is 0.54 which shows a modest degree of correlation between the two variables, indicating that those who see themselves as members of the leading crowd tend to see no compromising of their principles in belonging.

Although this interpretation appears almost obvious, in retrospect, it differs from Coleman's original analysis and conclusions. This example is a good illustration of the way that a judicious blend of statistical techniques and substantive insights can uncover the underlying structure of the data.

The application to the teaching styles data by Aitkin *et al.* (1981) involved 38 items and we shall not present the full analysis. We include it because a number of problems were encountered which are likely to occur in large investigations. The procedure was to fit a 2-class model using the E-M algorithm. This was successfully done and the authors were satisfied that the solution reached was a global maximum. The classes were estimated to be of similar size with $\hat{\eta}_1 = 0.538$. The estimates of π_{i0} ranged from 0.09 to 0.97 and those of π_{i1} from 0.09 to 0.95. In some cases $\hat{\pi}_{i0}$ and $\hat{\pi}_{i1}$ were similar and were thus of little use for naming the classes. The authors identified about 12 items where these two probabilities were substantially different, and on this basis it seemed reasonable to identify the first (zero) class with the more formal style of teaching.

Judging the adequacy of the model proved difficult. One approach was to test the hypothesis that the data had come from a single population against the alternative of a 2-class model. Although the likelihood ratio can be found, its asymptotic distribution is not that of

χ^2 because the null hypothesis is a boundary point of the parameter space. A simulation technique, due to Hope (1968), was used to show that the null hypothesis could be rejected. A second method was to construct a formality score by adding up the number of questions to which each individual gave the more "formal" answer. The shape of the distribution was then examined to see whether it was more consistent with a mixture of 2 or 3 normal distributions.

The method of comparing the observed and expected frequencies over the score patterns was not available. The 2^p contingency table has 2^{38} cells of which all but 468 were empty and no cell had more than one occupant.

Aitkin *et al.* (1981) went on to fit models with from 3 to 7 latent classes, but a second local maximum appeared with 3 classes and beyond that point there were multiple maxima. As noted earlier it is difficult to know how to interpret the maximum likelihood solution in such cases and the authors made no attempt to go further than the 3-class solution. In that case the 3 classes had estimated prior probabilities of $\hat{\eta}_1 = 0.366$, $\hat{\eta}_2 = 0.312$ and $\hat{\eta}_3 = 0.322$. The first two classes had much the same interpretation as before, but the third class appeared to have separated off a distinct group who had more difficulty with discipline problems and in other respects were intermediate between the two classes.

No standard errors were calculated, which means that some of the differences noted might have little significance but, overall, the 3-class hypothesis seems well supported.

Macready and Dayton's Mastery Model

Table 2.2 shows the results of a test on four items selected at random from a domain of items each involving the multiplication of a two-digit number by a three- or four-digit number involving "carry" operations. Full details, further examples and applications are given in Macready and Dayton (1977). According to the authors' model an individual is a "master" with probability η, say, and otherwise a non-master. In our notation, π_{i0} will be the probability that a non-master gets item i correct by guessing, and π_{i1} will be the probability that a master gets the same item correct; $1 - \pi_{i1}$ is therefore the "forgetting" probability. The estimates obtained by fitting the model by maximum likelihood, together with standard errors, are set out in Table 2.3.

It is clear from the goodness of fit test that the model provides a very good fit though, as Table 2.3 shows, the asymptotic standard errors are large. There is only clear evidence of guessing on item 1 and

Table 2.2 Macready and Dayton's data with posterior probabilities of belonging to the mastery state

Response pattern	Frequency	Expected frequency	$Pr\{\text{Master} \mid \mathbf{x}\}$
1111	15	15.0	1.00
1110	7	6.0	1.00
1101	23	19.7	1.00
1100	7	8.9	.91
1011	1	4.2	1.00
1010	3	1.9	.90
1001	6	6.1	.91
1000	13	12.9	.18
0111	4	4.9	1.00
0110	2	2.1	.97
0101	5	6.6	.98
0100	6	5.6	.47
0011	4	1.4	.97
0010	1	1.3	.43
0001	4	4.0	.45
0000	41	41.0	.02
	142	41.8	

$\chi^2 = 2.77$ with 3 degrees of freedom (after grouping to ensure expected frequencies greater than 5).

it appears that masters are most prone to forget item 3. Table 2.2 also gives the estimated probabilities that a person with a given response pattern has mastered the technique. Since we estimate that 41% are non-masters and since most of these must come from the 0000 group there are only about 17 ((.41)142 − 41) to be found from the other response categories. Hence anyone with only two items right is very likely to be a master, though note that if one of these two is item 1, which can be guessed correctly by a non-master with probability 0.21, the probability is somewhat less. Getting item 1 correct on its own provides very weak evidence of mastery for the same reason.

An extensive investigation of the applicability of this and related models is given in Wamani (1985).

Table 2.3 Parameter estimates for the fit of Macready and Dayton's model to the data of Table 2.2 (standard errors in brackets)

j	$\hat{\pi}_{1j}$	$\hat{\pi}_{2j}$	$\hat{\pi}_{3j}$	$\hat{\pi}_{4j}$	$\hat{\eta}$
0	.21 (.07)	.07 (.06)	.02 (.03)	.05 (.05)	.59 (.06)
1	.75 (.06)	.78 (.06)	.43 (.06)	.71 (.07)	

2.3 Latent class models with polytomous manifest variables

The extension of the foregoing theory to polytomous data is straightforward though applications have been less common. Some theory of estimation was given by Goodman (1978) and we shall describe applications by Pickering and Forbes (1984) and by Clogg (1979) (re-analysed by Masters (1985)).

When there are more than two categories the indicator variable x_i used previously is replaced by a vector $\mathbf{x}_i$ with c_i elements $x_i(s)$ defined by

$$x_i(s) = 1 \text{ if the response is in category } s \text{ on variable } i$$
$$= 0 \text{ otherwise} \quad (s = 0, 1, \ldots, c_i - 1).$$

Note that $\sum_s x_i(s) = 1$. The complete response vector for an individual is then written

$$\mathbf{x}' = (\mathbf{x}_1', \mathbf{x}_2', \ldots, \mathbf{x}_p').$$

(This notation will be used again in Chapter 6.) The conditional response probabilities are defined by

$\pi_{ij}(s) = Pr\{$Response of an individual in class j is in category s on variable $i\}$.

The joint probability function of $\mathbf{x}$ is then

$$f(\mathbf{x}) = \sum_{j=0}^{K-1} \eta_j \prod_{i=1}^{p} \prod_{s=1}^{c_i-1} \{\pi_{ij}(s)\}^{x_i(s)}. \tag{2.14}$$

The posterior distribution is

$$h(j \mid \mathbf{x}) = \eta_j \prod_{i=1}^{p} \prod_{s=1}^{c_i-1} \{\pi_{ij}(s)\}^{x_i(s)} / f(\mathbf{x}). \tag{2.15}$$

Maximum Likelihood Estimation

As with binary data, the log-likelihood may be written

$$L = \sum_{h=1}^{n} \log f(\mathbf{x}_h)$$

but the maximization must now be effected under two sets of constraints. The original constraint $\sum_j \eta_j = 1$ remains, but we must now impose

$$\sum_{s=0}^{c_i-1} \pi_{ij}(s) = 1 \quad (i = 1, 2, \ldots, p). \tag{2.16}$$

This did not arise in the binary case because we used (2.16) to eliminate one of the two probabilities for each dimension. The function now to be maximized is thus

$$\phi = L + \theta \sum_{j=0}^{K-1} \eta_j + \sum_{j=0}^{K-1} \sum_{i=1}^{p} \beta_{ij} \sum_{s=0}^{c_i - 1} \pi_{ij}(s) \tag{2.17}$$

where θ and $\{\beta_{ij}\}$ are undetermined multipliers. The partial derivatives with respect to $\{\eta_j\}$ are

$$\frac{\partial \phi}{\partial \eta_j} = \sum_{h=1}^{n} f(\mathbf{x}_h \mid j)/f(\mathbf{x}_h) + \theta \tag{2.18}$$

leading, as before, to

$$\hat{\eta}_j = \frac{1}{n} \sum_{h=1}^{n} h(j \mid \mathbf{x}_h). \tag{2.19}$$

Similarly

$$\frac{\partial L}{\partial \pi_{ij}(s)} = \sum_{h=1}^{n} \eta_j \frac{\partial f(\mathbf{x}_h \mid j)}{\partial \pi_{ij}(s)} \Big/ f(\mathbf{x}_h).$$

Now

$$\begin{aligned} \frac{\partial f(\mathbf{x}_h \mid j)}{\partial \pi_{ij}(s)} &= \frac{\partial}{\partial \pi_{ij}(s)} \exp \sum_{i=1}^{p} \sum_{s=0}^{c_i - 1} x_{ih}(s) \log \pi_{ij}(s) \\ &= f(\mathbf{x}_h \mid j) x_{ih}(s)/\pi_{ij}(s) \end{aligned}$$

giving

$$\begin{aligned} \frac{\partial \phi}{\partial \pi_{ij}(s)} &= \eta_j \sum_{h=1}^{n} \frac{f(\mathbf{x}_h \mid j)}{f(\mathbf{x}_h)} \frac{x_{ih}(s)}{\pi_{ij}(s)} + \beta_{ij} \\ &= \sum_{h=1}^{n} h(j \mid \mathbf{x}_h) x_{ih}(s)/\pi_{ij}(s) + \beta_{ij}. \end{aligned} \tag{2.20}$$

Setting the right-hand side equal to zero yields

$$\pi_{ij}(s) = \sum_{h=1}^{n} h(j \mid \mathbf{x}_h) x_{ih}(s) + \pi_{ij}(s) \beta_{ij}. \tag{2.21}$$

Summing both sides over s

$$1 = \sum_{h=1}^{n} h(j \mid \mathbf{x}_h) + \beta_{ij} \quad \text{or} \quad \beta_{ij} = 1 - \sum_{h=1}^{n} h(j \mid \mathbf{x}_h).$$

Substitution into (2.21) finally gives the estimating equations

$$\hat{\pi}_{ij}(s) = \sum_{h=1}^{n} h(j \mid \mathbf{x}_h) x_{ih}(s) \Big/ \sum_{h=1}^{n} h(j \mid \mathbf{x}_h)$$

$$= \sum_{h=1}^{n} h(j \mid \mathbf{x}_h) x_{ih}(s) / n\hat{\eta}_j \quad (i = 1, 2, \ldots, p; j = 0, 1, \ldots, K-1). \tag{2.22}$$

The equations (2.19) and (2.22) may then be solved as in the binary E-M algorithm. We first choose starting values for $\{\hat{h}(j \mid \mathbf{x}_h)\}$ and then obtain first approximations to $\{\hat{\eta}_j\}$ and $\{\hat{\pi}_{ij}(s)\}$. These are then used to improve the estimates of $\{\hat{h}(j \mid \mathbf{x}_h)\}$ and so on.

Allocation of Individuals to Latent Classes

Reference to (2.15) shows that

$$h(j \mid \mathbf{x}_h)/h(k \mid \mathbf{x}_h) = (\eta_j/\eta_k) \exp \sum_{i=1}^{p} \sum_{s=0}^{c_i-1} x_i(s) \log \pi_{ij}(s)/\pi_{ik}(s).$$

Discrimination is thus based on comparing the linear functions

$$X_j = \sum_{i=1}^{p} \sum_{s=0}^{c_i-1} x_i(s) \log \pi_{ij}(s) \quad (j = 0, 1, \ldots, K-1).$$

Thus class j is preferred to class k if

$$X_j - X_k > \log(\eta_k/\eta_j). \tag{2.23}$$

This implies that we allocate to the class for which

$$X_j + \log \eta_j$$

is greatest.

Interpretation

There is little here to add to the discussion of the binary case. Identification of the classes is made by reference to the set of responses which are either common or rare for members of that class.

Asymptotic standard errors can be computed in the usual way from the second derivatives which are easily obtained in the same manner as in the binary case. Goodness of fit can be tested by comparing observed and expected frequencies of response patterns in the usual way, but the problems arising from sparsely occupied cells when p is large are even more acute than in the binary case.

Classification of Scottish infants

As our first example we return to the classification of Scottish infants investigated by Pickering and Forbes (1984). This involved an

Table 2.4 Parameter estimates for the 2- and 4-Class models using 1980 data and complete cases only

Variable	No. of levels	Levels	2 Classes		3 Classes			4 Classes			
Birthweight	4	2001–2500 g	0.01	0.48	0.00	0.18	0.79	0.00	0.20	0.78	0.03
		1501–2000 g	0.00	0.15	0.00	0.26	0.09	0.00	0.32	0.09	0.00
		⩽1500 g	0.00	0.08	0.00	0.21	0.01	0.00	0.25	0.01	0.00
Birthweight for gestation age	2	<10th centile	0.07	0.43	0.07	0.19	0.62	0.07	0.21	0.62	0.10
Apgar at 5 min	2	<7	0.01	0.12	0.01	0.26	0.01	0.00	0.21	0.01	0.32
Resuscitation	3	Intermediate	0.08	0.19	0.08	0.26	0.13	0.07	0.25	0.13	0.31
		By intubation	0.02	0.17	0.02	0.33	0.04	0.01	0.29	0.03	0.52
Assisted Ventilation after 30 mins	2	Present	0.00	0.10	0.00	0.25	0.00	0.00	0.29	0.00	0.01
Recurrent apnoea	2	Present	0.00	0.06	0.00	0.17	0.00	0.00	0.20	0.00	0.00
Jaundice	2	Present	0.28	0.58	0.28	0.67	0.49	0.28	0.71	0.49	0.32
Convulsions	2	Present	0.00	0.03	0.00	0.07	0.00	0.00	0.07	0.00	0.01
In tube feeding	2	Present	0.01	0.30	0.01	0.60	0.10	0.01	0.67	0.10	0.05
Dead at discharge	2	Present	0.00	0.05	0.00	0.13	0.00	0.00	0.15	0.00	0.00
Age at discharge	3	4–10 days	0.84	0.34	0.83	0.09	0.53	0.83	0.04	0.50	0.84
		>11 days	0.03	0.61	0.03	0.79	0.45	0.03	0.82	0.46	0.14
Relative frequency of class ($\hat{\eta}_j$)			0.92	0.08	0.92	0.03	0.05	0.89	0.03	0.05	0.04

unusually large sample size (45,426 usable cases relating to the year 1980) which permitted some interesting additions to the usual analysis. There were 11 variables of which 8 were binary; they are listed in Table 2.4 which also gives the parameter estimates obtained using the E-M algorithm. It is only necessary to give the estimates for $(c_i - 1)$ categories of each variable because the response probabilities for each variable sum to one. There are the same difficulties about testing goodness of fit here as in the teaching styles example. Although the value of χ^2 continues to decrease as the number of classes fitted increases, multiple maxima began to appear when 4 classes were fitted. The decision to present results up to the 4-class model was based on the interpretability of the results rather than on statistical criteria. However, with such a large sample it was possible to test the stability of the results by dividing the sample in half at random and re-fitting the model to each sub-sample. The authors showed that the parameter estimates obtained were virtually the same. The 4-class solution distinguishes between a large class of normal healthy infants (class I) and three additional classes of infants whose births are abnormal in some way. Two classes (III and IV) represent moderately ill infants requiring special neonatal care and the third (II) represents severely ill infants with low birthweight.

Using these results one can predict the prevalence of each class and study their geographical distribution.

Life Satisfaction Data

The second example is taken from Clogg (1979) who fitted a 3-class latent variable model to data from the 1975 US General Household Survey with a sample size of 1472. The data were re-analysed by Masters (1985) using a latent trait model (see Chapter 9).

The data concern three questions about degree of satisfaction with family (F), hobbies (H) and residence (R). Answers to each question were classified as low, medium or high. It is reasonable to suppose that if individuals tend to give the same reply to each question then the distribution might be explained by a model with three classes and this is what Clogg fitted. The method used was maximum likelihood using Goodman's E-M routine. The parameter estimates are given in Table 2.5 and the observed and expected frequencies in Table 2.6.

The goodness of fit may be judged by the value of $\chi^2 = 2.36$ which with 6 degrees of freedom $(27 - 18 - 2 - 1)$ indicates an extremely good fit.

We note from the $\hat{\eta}$'s in Table 2.5 that the first class is very small, but the pattern of the π's confirms our expectation that there might be a tendency to give similar answers to all questions. Thus those in class

Table 2.5 Parameter estimates (π's) of 3-class latent variable models for life satisfaction data

Class		I	II	III
Family	low	.48	.06	.03
	medium	.28	.36	.02
	high	.24	.58	.95
Hobbies	low	.85	.18	.07
	medium	.00	.50	.14
	high	.15	.32	.79
Residence	low	.53	.21	.06
	medium	.31	.53	.22
	high	.16	.26	.72
	$\hat{\eta}_j$	.04	.41	.55

I tend to have low satisfaction, those in class II somewhat higher satisfaction and those in class III highest of all. Indeed the choice of questions with three alternatives may well have divided the respondents into three classes by requiring them to think about life satisfaction in this way. To that extent the excellent fit may be, in part, an artefact, even though responses to all questions reflect a basic degree of satisfaction.

Table 2.6 Observed and expected frequencies for life satisfaction data ($n = 1472$)

Response pattern F	H	R	Observed frequency	Expected frequency	Response pattern F	H	R	Observed frequency	Expected frequency
0	0	0	15	14.1	2	1	0	45	43.3
0	0	1	11	11.3	2	1	1	117	116.4
0	0	2	7	6.8	2	1	2	126	123.0
1	0	0	16	15.6	0	2	0	5	5.7
1	0	1	26	25.3	0	2	1	14	11.7
1	0	2	12	13.2	0	2	2	16	17.5
2	0	0	23	22.8	1	2	0	18	16.6
2	0	1	40	48.9	1	2	1	38	40.4
2	0	2	54	57.0	1	2	2	27	27.8
0	1	0	3	4.0	2	2	0	64	61.2
0	1	1	12	10.3	2	2	1	191	193.6
0	1	2	5	7.2	2	2	2	466	466.9
1	1	0	23	23.0		Total		1472	1472.0
1	1	1	58	58.1					
1	1	2	31	29.9					

Individuals may be allocated to classes using the estimated parameters and the allocation rule of Section 2.3. The result is that those with response patterns 000, 001 and 002 go into class I, those with response patterns 222, 221, 220, 212, 202, 022, into class III and the remainder into class II. Membership of class I is thus determined by low satisfaction with family and hobbies; the make-up of class III is less transparent but high satisfaction in at least two areas is the main qualification.

2.4 Latent class models with metrical manifest variables

At the beginning of Section 1.4 we gave an example of a latent class model with metrical data and we noted that the use of such a model would require the choice of a suitable form for the conditional distribution of $\{x_i\}$. When dealing with binary or polytomous data the binomial and multinomial distributions were the obvious choice, but with metrical data there may be some difficulty in knowing what to assume. Some information can be gleaned from the marginal distributions as we note below. In this section we are concerned with a model where the joint distribution of the x's has the form

$$f(\mathbf{x}) = \sum_{j=0}^{K-1} \eta_j \prod_{i=1}^{p} g_i(x_i \mid j) \tag{2.24}$$

for some K, where $g_i(x_i \mid j)$ is the conditional distribution of x_i for members of class j. Such models have been termed latent profile models.

In the absence of information on the form of $g_i(x_i \mid j)$ some guidance may be obtained from an inspection of the marginal distributions given by

$$f(x_i) = \sum_{j=0}^{K-1} \eta_j g_i(x_i \mid j) \quad (i = 1, 2, \ldots, p). \tag{2.25}$$

A bimodal distribution would suggest a 2-class normal mixture whereas an extremely skew distribution would argue against a normal mixture of any number of components. At best we are only likely to obtain very crude information in this way, but it should be possible to avoid grossly inappropriate assumptions. In favourable circumstances it may be possible to use the marginal distributions to distinguish latent profile models from the factor models to be considered in Chapter 3, as we shall see below. One can, of course, fit mixture models to the marginal distributions by methods given in Everitt and Hand (1981) but this would not solve our present problem unless the η's were constrained to be the same for each margin. We shall describe two approaches to fitting the model of (2.24), the second of which throws some interesting light on the interpretation of covariance structures.

Maximum Likelihood Estimation

The same approach can be used as with categorical manifest variables, though the details depend on our choice of $\{g_i(x_i \mid j)\}$. The equations obtained by setting the derivatives of L with respect to $\boldsymbol{\eta}$ equal to zero are the same for all choices including the distributions used for categorical data and thus are

$$\hat{\eta}_j = \sum_{h=1}^{n} h(j \mid \mathbf{x}_h)/n \quad (j = 0, 1, \ldots, K-1). \tag{2.26}$$

Suppose we now take

$$g_i(x_i \mid j) \equiv g(x_i \mid \theta_{ij})$$

then

$$L = \sum_{h=1}^{n} \log \sum_{j=0}^{K-1} \eta_j g(x_{ih} \mid \theta_{ij}) \tag{2.27}$$

and

$$\frac{\partial L}{\partial \theta_{ij}} = \sum_{h=1}^{n} \eta_j \frac{\partial g}{\partial \theta_{ij}} \bigg/ g(x_{ih} \mid \theta_{ij}). \tag{2.28}$$

Setting these derivatives equal to zero and solving for $\{\theta_{ij}\}$ we shall have equations of the form

$$\theta_{ij} = \psi(\mathbf{x}_h, \boldsymbol{\eta}) \quad (i = 1, 2, \ldots, p; j = 0, 1, \ldots, K-1). \tag{2.29}$$

The E-M procedure can then be used, as before, by alternating between (2.26) and (2.29). The second derivatives can be used to obtain asymptotic standard errors.

The equations (2.29) take a particularly simple form if $g(x_i \mid \theta_{ij})$ is a member of the exponential family and especially if it is normal with mean θ_{ij} and unit variance. We then find

$$\frac{\partial g}{\partial \theta_{ij}} = (x_{ih} - \theta_{ij}) g(x_i \mid \theta_{ij})$$

and hence that

$$\sum_{h=1}^{n} h(j \mid \mathbf{x}_h)(x_{ih} - \theta_{ij}) = 0$$

or

$$\hat{\theta}_{ij} = \sum_{h=1}^{n} x_{ih} h(j \mid \mathbf{x}_h) \bigg/ \sum_{h=1}^{n} h(j \mid \mathbf{x}_h). \tag{2.30}$$

Equations (2.30) and (2.26) then lend themselves to a straightforward application of the E-M procedure.

Other Methods

The original method of estimation proposed for the latent profile model by Lazarsfeld and Henry (1968) involved fitting by the method of moments. In this case one writes down sufficient moment, and cross-moment, equations to determine the unknown parameters. As a method of estimation this has been superseded by the maximum likelihood approach, but it brings out an important link with the factor analysis model which is not otherwise evident.

Let $\mu_i(j)$ be the mean of x_i for members of latent class j and let $\sigma_i^2(j)$ be its variance. Then

$$E(x_i) = \sum_{j=0}^{K-1} \eta_j \int \cdots \int x_i \prod_{i=1}^{p} g_i(x_i \mid j)\, d\mathbf{x} = \sum_{j=0}^{K-1} \eta_j \mu_i(j) \tag{2.31}$$

$$E(x_i^2) = \sum_{j=0}^{K-1} \eta_j \{\sigma_i^2(j) + \mu_i^2(j)\} \tag{2.32}$$

$$E(x_i x_k) = \sum_{j=0}^{K-1} \eta_h \mu_i(j) \mu_k(j) \quad (i, k = 1, 2, \ldots, p; i \neq j). \tag{2.33}$$

We then have

$$\text{var}(x_i) = \sum_{j=0}^{K-1} \eta_j \sigma_i^2(j) + \sum_{j=0}^{K-1} \eta_j \{\mu_i(j) - \bar{\mu}_i\}^2, \quad (i = 1, 2, \ldots, p) \tag{2.34}$$

$$\begin{aligned} \text{cov}(x_i, x_k) &= \sum_{j=0}^{K-1} \eta_j \mu_i(j) \mu_k(j) - \left\{ \sum_{j=0}^{K-1} \eta_j \mu_i(j) \right\}^2 \\ &= \sum_{j=0}^{K-1} \eta_j \{\mu_i(j) - \bar{\mu}_i\}\{\mu_k(j) - \bar{\mu}_k\} \\ &\qquad\qquad (i, k = 1, 2, \ldots, p; i \neq k) \end{aligned} \tag{2.35}$$

where $\bar{\mu}_i = \sum_{j=0}^{K-1} \eta_j \mu_i(j)$. The dispersion matrix may thus be written

$$D(\mathbf{x}) = \mathbf{L}\mathbf{L}' + \boldsymbol{\psi} \tag{2.36}$$

where $\boldsymbol{\psi}$ is now a diagonal matrix with (i, i)th element $\sum_{j=0}^{K-1} \eta_j \sigma_i^2(j)$ and the element l_{ij} of $\mathbf{L}$ is given by

$$l_{ij} = \sqrt{\eta_j} \{\mu_i(j) - \bar{\mu}_i\}. \tag{2.37}$$

Thus $D(\mathbf{x})$ is of exactly the same form as the dispersion matrix for the linear factor model with a normally distributed latent variable, but there is one important difference. The columns of $\mathbf{L}$ are linearly dependent because

$$\sum_{j=0}^{K-1} \sqrt{\eta_j} l_{ij} = 0 \quad (i = 1, 2, \ldots, p)$$

from the definition of $\bar{\mu}_i$. For this reason any attempt to estimate $\mathbf{L}$ and $\boldsymbol{\psi}$ by fitting the sample dispersion matrix to (2.36) by standard methods (for which see Chapter 3) would fail. However, there exists a $p \times (K-1)$ matrix $\boldsymbol{\Lambda}$ with linearly independent columns such that $\mathbf{LL}' = \boldsymbol{\Lambda\Lambda}'$ (see Graybill (1983) Theorem 1.7.7). We can then write

$$D(\mathbf{x}) = \boldsymbol{\Lambda\Lambda}' + \boldsymbol{\psi} \tag{2.38}$$

which makes the correspondence complete.

This result has two very important practical implications. One is that the methods of fitting to be given in Chapter 3 could be used to estimate $\boldsymbol{\Lambda}$ and $\boldsymbol{\psi}$ for this model. (But note that this would only give us the scaled deviations of (2.37) and the weighted conditional variances.) The second implication is that the factor model and the latent profile models are indistinguishable so far as their covariance structure is concerned. Thus one could successfully fit a standard linear factor model by covariance methods, even though the data had arisen from a latent class model. This was one reason for emphasizing the desirability of examining the marginal distributions since multimodality there would be one way of distinguishing a latent class model from the factor model. Unfortunately, mixtures of distributions do not always exhibit distinct modes, so an inspection of the margins may leave the matter unresolved.

The difficulty of distinguishing these models is an example of the general point that the form of the distribution of latent variables is poorly determined by the data. In Chapter 4 we shall gain some insight into why this is so, and other examples will be encountered in the course of the book. In the present case we are saying that a latent variable with a continuous normal distribution and one in which all the prior probability is concentrated at K points lead to very similar data matrices.

The relationship between $\boldsymbol{\Lambda}$ and $\mathbf{L}$ is complicated in general, but when $K = 2$ it is both simple and informative. We first note that

$$\mu_i(1) - \bar{\mu}_i = (1 - \eta_1)\{\mu_i(1) - \mu_i(2)\}$$
$$\mu_i(2) - \bar{\mu}_i = -\eta_1\{\mu_i(1) - \mu_i(2)\}.$$

$\mathbf{L}$ is thus a matrix with ith row equal to

$$[\sqrt{\eta_1}(1 - \eta_1)\{\mu_i(1) - \mu_i(2)\}, \quad -\sqrt{\eta_2}\eta_1\{\mu_i(1) - \mu_i(2)\}].$$

The (i, k)th element of $\mathbf{LL}'$ is thus

$$\eta_1(1-\eta_1)\{\mu_i(1)-\mu_i(2)\}\{\mu_k(1)-\mu_k(2)\}$$

and this is identical to the corresponding element in $\mathbf{\Lambda\Lambda}'$ when $\mathbf{\Lambda}$ is a $p \times 1$ matrix with ith element

$$\sqrt{\eta_1(1-\eta_1)}\{\mu_i(1)-\mu_i(2)\}. \tag{2.39}$$

The same result could have been obtained by starting with the linear model of (1.11), setting $q=1$ and allowing y to have the Bernoulli distribution with

$$Pr\{y=1\}=\eta_1.$$

Looking at the problem in this way shows why the case $K>2$ is less simple since then we need a vector-valued indicator for $\mathbf{y}$. We may also note that the covariance structure of the latent profile model will only yield information about the means of the conditional distributions of $\{x_i\}$. ($\boldsymbol{\psi}$ depends on the variances but only in a weighted average form).

Allocation to Categories

We can proceed as in the latent class model by calculating the posterior probabilities of class membership and allocating accordingly. Thus

$$h(j \mid \mathbf{x})=\eta_j f(\mathbf{x} \mid j)/f(\mathbf{x})$$

and therefore class j is more probable than class k if

$$\frac{\eta_j}{\eta_k}\frac{f(\mathbf{x} \mid j)}{f(\mathbf{x} \mid k)}>1.$$

If the x's are normal so that

$$f(\mathbf{x} \mid j)=(2\pi)^{-\frac{1}{2}p}\exp-\tfrac{1}{2}\sum_{i=1}^{p}(x_i-\mu_i(j))^2$$

and hence

$$f(\mathbf{x} \mid j)/f(\mathbf{x} \mid k)=\exp\left\{\sum_{i=1}^{p}x_i\mu_i(j)-\tfrac{1}{2}\sum_{i=1}^{p}\mu_i^2(j)-\sum_{i=1}^{p}x_i\mu_i(k)+\tfrac{1}{2}\sum_{i=1}^{p}\mu_i^2(k)\right\}$$

the allocation rule is then made on the basis of the value of

$$\sum_{i=1}^{p}x_i\mu_i(j)-\tfrac{1}{2}\sum_{i=1}^{p}\mu_i^2(j)+\log\eta_j. \tag{2.40}$$

Once again the rule for allocation turns out to be based on a linear function of the data. This will be a recurring feature of many latent variable models which arises whenever the conditional distributions involved belong to the exponential family.

CHAPTER 3

The Linear Factor Model

3.1 Definition and basic properties

In Section 1.4 we arrived at a factor model with two rather special properties. As we see from (1.11) the y's affect only the mean of the conditional distribution of the x's and that mean is a linear function of the y's. These two properties define what we here term a *linear factor model.* If we introduce the further assumptions of normality as set out in (1.11) we shall have a *normal linear factor model.* Other distributional assumptions would lead to other members of the family but all would share the two defining features, namely that

(a) $E(\mathbf{x} \mid \mathbf{y}) = \boldsymbol{\mu} + \boldsymbol{\Lambda}\mathbf{y}$

and

(b) the conditional distribution of $\mathbf{x}$ given $\mathbf{y}$ depends on $\mathbf{y}$ only through its mean.

An equivalent way of expressing (a) and (b) is to write

$$\mathbf{x} = \boldsymbol{\mu} + \boldsymbol{\Lambda}\mathbf{y} + \mathbf{e} \tag{3.1}$$

where $\mathbf{e}$ is a p-vector of independent random variables with $E(\mathbf{e}) = \mathbf{0}$. Because the x's are conditionally independent it follows that the e's are independent and uncorrelated with the y's.

Equation (3.1) is the usual starting-point for the development of factor analysis and this is the path that we shall follow in this chapter. However, it is worth reiterating two points about (3.1) which have already emerged in Chapter 1 but are not often commented upon. The first is that the assumption that the elements of $\mathbf{e}$ are independent is not an assumption in the usual sense. It is an alternative way of expressing the postulate of conditional independence. Although it is easily overlooked, q (the dimension of $\mathbf{y}$), is a parameter of the model and the aim is to find the smallest q for which the $\mathbf{e}$'s have the properties required by the model. The second point is that there are other models which are equivalent to (3.1). This arbitrariness has two aspects, one of which is well-recognized and to which we shall come below. The other arises from the point made in Section 1.4 about the effect of transforming the y's. To make the point again by an example, consider the model

$$\mathbf{x} = \boldsymbol{\mu} + \boldsymbol{\Lambda}\mathbf{u} + \mathbf{e} \tag{3.2}$$

where $\mathbf{u} = (\log v_1, \log v_2, \ldots, \log v_q)$; this is equivalent to the normal version of (3.1) if the v's have lognormal distributions. There can thus be no empirical ground for regarding the u's as a "better" way of specifying the factors than the v's. If, for example, we had a model with one factor, there is nothing the data can tell us about what its prior distribution really is. Any choice we make must therefore be regarded as no more than a convenient convention.

The second type of arbitrariness is an extension of that arising from orthogonal transformations of the latent variables. More generally, consider

$$\mathbf{x} = \boldsymbol{\mu} + \boldsymbol{\Lambda}\mathbf{M}^{-1}\mathbf{z} + \mathbf{e} \tag{3.3}$$

where $\mathbf{z} = \mathbf{M}\mathbf{y}$ and $\mathbf{M}$ is a $q \times q$ non-singular matrix. This model is equivalent to that of (3.1) yet it has weight matrix $\boldsymbol{\Lambda}\mathbf{M}^{-1}$ instead of $\boldsymbol{\Lambda}$ and its factors are linear transformations of the original y's. The z's, of course, will not, in general, be independent.

The three types of model represented by (3.1), (3.2) and (3.3) are best regarded as one model with the differing forms of the right-hand sides indicating different ways of describing the same situation.

Fitting the linear model is usually approached via the dispersion matrix of the x's, here written $D(\mathbf{X}) = \boldsymbol{\Sigma}$. Without loss of generality we may assume the y's to be uncorrelated and have unit variance. It then follows that

$$\boldsymbol{\Sigma} = \boldsymbol{\Lambda}\boldsymbol{\Lambda}' + \boldsymbol{\psi} \tag{3.4}$$

where $\boldsymbol{\psi}$ is a diagonal matrix with $\psi_i = \text{var}(e_i)$, $(i = 1, 2, \ldots, p)$. If we assume a *normal* linear model the joint distribution of the x's is multivariate normal with dispersion matrix (3.4). In the general case we aim to fit the model by choosing estimators so that the observed and theoretical dispersion matrices are as close as possible in some sense.

Before leaving these preliminaries we give an heuristic argument for believing that results derived for the normal linear factor model will be robust with respect to departures from normality. Estimates and their sampling properties depend on the distribution of the x's and not, directly, on those of the y's or the e's. Since, by (3.1), the x's are linear functions of independent random variables with finite variances, the central limit theorem will ensure that the x's are approximately normal, especially if q is moderately large. Even in the most unfavourable case, when $q = 1$, each x is still a linear combination of two independent random variables, and unless either is highly non-normal the x's may be expected to be roughly normal. For this reason we would not expect the individual distributions of the y's or the e's to have very much influence on the joint distribution of the x's. We shall find confirmation of this conclusion later but, in the meantime, it shows why the results for the normal model now to be discussed have wider applicability than might appear at first sight.

3.2 The normal linear factor model

This arises when the y's and e's have independent normal distributions. It may be formally specified as in (1.1) or by saying that in (3.1) $y \frown N_q(\mathbf{0}, \mathbf{I})$ and $\mathbf{e} \frown N_p(\mathbf{0}, \boldsymbol{\psi})$. In this section we shall discuss how to estimate $\boldsymbol{\Lambda}$, $\boldsymbol{\psi}$ and q. The interpretation will come later in Section 3.4. We use the method of maximum likelihood which is straightforward and yields an asymptotic test of fit and standard errors.

We saw in (1.12) that the joint distribution of the x's is

$$\begin{aligned} f(\mathbf{x}) &= (2\pi)^{-p/2} |\boldsymbol{\Sigma}|^{-\frac{1}{2}} \exp -\tfrac{1}{2}(\mathbf{x}-\boldsymbol{\mu})'\boldsymbol{\Sigma}^{-1}(\mathbf{x}-\boldsymbol{\mu}) \\ &= (2\pi)^{-p/2} |\boldsymbol{\Sigma}|^{-\frac{1}{2}} \exp -\tfrac{1}{2} \operatorname{tr} \boldsymbol{\Sigma}^{-1}(\mathbf{x}-\boldsymbol{\mu})(\mathbf{x}-\boldsymbol{\mu})' \end{aligned} \tag{3.5}$$

where $\boldsymbol{\Sigma} = \boldsymbol{\Lambda}\boldsymbol{\Lambda}' + \boldsymbol{\psi}$. The log-likelihood for a random sample of size n is thus

$$L = -\tfrac{1}{2}np \log 2\pi - \tfrac{1}{2}n \log|\boldsymbol{\Sigma}| - \tfrac{1}{2}n \operatorname{tr} \boldsymbol{\Sigma}^{-1}\mathbf{S} \tag{3.6}$$

where

$$\mathbf{S} = \frac{1}{n}\sum_{h=1}^{n} (\mathbf{x}_h - \boldsymbol{\mu})(\mathbf{x}_h - \boldsymbol{\mu})'. \tag{3.7}$$

Maximum Likelihood Estimation

We obtain the estimators in the usual way by setting the partial derivatives equal to zero. This is not strictly correct since the elements of $\boldsymbol{\psi}$ are restricted to be non-negative and hence the absolute maximum of L may occur outside the parameter space. The maximum we seek will then occur on a boundary on which at least one of the ψ's will be zero and for which the corresponding partial derivative will not be zero. Such cases, known as Heywood cases, occur quite often and will be discussed further in Section 3.6. For the present we note that if the absolute maximum of (3.6) lies within the admissible parameter space then it is the solution we are seeking. We therefore address ourselves to the problem of finding a point at which the partial derivatives vanish. We must also remember in this connection that the value of $\boldsymbol{\Sigma}$, and hence of the likelihood, is unchanged by an orthogonal transformation of $\mathbf{\Lambda}$ to $\mathbf{\Lambda M}'$. The likelihood will not therefore have a single maximum but infinitely many. However, if we can find any one point at which L is a maximum all others can be found by orthogonal transformations. We shall be able to use this fact to simplify the solution of the likelihood equations.

Differentiating (3.6) with respect to $\boldsymbol{\mu}$ yields the estimators $\boldsymbol{\mu} = \bar{\mathbf{x}}$ in the usual way. To economize on notation we shall henceforth use $\mathbf{S}$ to denote the matrix of (3.7) with $\boldsymbol{\mu}$ replaced by $\hat{\boldsymbol{\mu}}$.

To find the partial derivatives of L with respect to $\mathbf{\Lambda}$ and $\boldsymbol{\psi}$ we first find those with respect to $\mathbf{\Sigma}$ and then relate them to those required. We find

$$\frac{\partial \log|\mathbf{\Sigma}|}{\partial \mathbf{\Sigma}} = 2\mathbf{\Sigma}^{-1} - \operatorname{diag}(\mathbf{\Sigma}^{-1})$$

and

$$\frac{\partial \operatorname{tr}(\mathbf{\Sigma}^{-1}\mathbf{S})}{\partial \mathbf{\Sigma}} = -2\mathbf{\Sigma}^{-1}\mathbf{S}\mathbf{\Sigma}^{-1} + \operatorname{diag}(\mathbf{\Sigma}^{-1}\mathbf{S}\mathbf{\Sigma}^{-1})$$

giving

$$-\frac{2}{n}\frac{\partial L}{\partial \mathbf{\Sigma}} = 2\mathbf{\Omega} - \operatorname{diag}\mathbf{\Omega} \tag{3.8}$$

where $\mathbf{\Omega} = \mathbf{\Sigma}^{-1}(\mathbf{\Sigma} - \mathbf{S})\mathbf{\Sigma}^{-1}$.

We have used here the notation of matrix differentiation. The derivative of a scalar, for example, L with respect to a matrix $\mathbf{\Sigma}$, say, is written

$$\frac{\partial L}{\partial \mathbf{\Sigma}}$$

and is identical to the matrix $\left\{\dfrac{\partial L}{\partial \sigma_{ij}}\right\}$. In more advanced work it is useful to have a matrix differential calculus as given, for example, in Magnus and Neudecker (1986). Now

$$\frac{\partial L}{\partial \lambda_{ij}} = \sum_{h=1}^{p} \left(\frac{\partial L}{\partial \sigma_{ih}}\right)\left(\frac{\partial \sigma_{ih}}{\partial \lambda_{ij}}\right)$$

where

$$\sigma_{ih} = \sum_{r=1}^{q} \lambda_{ir}\lambda_{hr} + \delta_{ih}\psi_i$$

is the (i, h)th element of $\mathbf{\Sigma}$ and δ_{ih} is the Kronecker delta. It follows that

$$\frac{\partial \sigma_{ih}}{\partial \lambda_{ij}} = (1 + \delta_{ih})\lambda_{hj}$$

and hence that

$$\frac{\partial L}{\partial \mathbf{\Lambda}} = \left\{\frac{\partial L}{\partial \mathbf{\Sigma}} + \text{diag}\,\frac{\partial L}{\partial \mathbf{\Sigma}}\right\}\mathbf{\Lambda}.$$

Substituting from (3.8), we have

$$\begin{aligned} -\frac{2}{n}\frac{\partial L}{\partial \mathbf{\Lambda}} &= \{2\mathbf{\Omega} - \text{diag}\,\mathbf{\Omega} + \text{diag}(2\mathbf{\Omega} - \text{diag}\,\mathbf{\Omega})\}\mathbf{\Lambda} \\ &= 2\mathbf{\Omega}\mathbf{\Lambda}. \end{aligned} \tag{3.9}$$

Similarly,

$$\frac{\partial L}{\partial \psi_i} = \frac{\partial L}{\partial \sigma_{ii}}\frac{\partial \sigma_{ii}}{\partial \psi_i}$$

and since

$$\frac{\partial \sigma_{ii}}{\partial \psi_i} = 1$$

$$-\frac{2}{n}\,\text{diag}\,\frac{\partial L}{\partial \boldsymbol{\psi}} = \text{diag}\,\mathbf{\Omega}. \tag{3.10}$$

In principle, the problem is now solved because, by setting (3.9) and (3.10) equal to zero there are sufficient equations to determine the unknowns. In practice the solution of these equations is not at all straightforward, but progress can be made if we first express them in an alternative form. We start with the equation obtained by setting (3.9) equal to zero. Pre-multiplication by $\mathbf{\Sigma}$ gives

$$(\mathbf{\Sigma} - \mathbf{S})\mathbf{\Sigma}^{-1}\mathbf{\Lambda} = \mathbf{0}. \tag{3.11}$$

This must be expressed wholly in terms of $\mathbf{\Lambda}$ and $\boldsymbol{\psi}$ which we do using the fact that

$$\mathbf{\Sigma}^{-1} = \boldsymbol{\psi}^{-1} - \boldsymbol{\psi}^{-1}\mathbf{\Lambda}(\mathbf{I} + \mathbf{\Gamma})^{-1}\mathbf{\Lambda}'\boldsymbol{\psi}^{-1} \tag{3.12}$$

where $\mathbf{\Gamma} = \mathbf{\Lambda}'\boldsymbol{\psi}^{-1}\mathbf{\Lambda}$. To verify this result we post-multiply both sides by $\mathbf{\Sigma} = \mathbf{\Lambda}\mathbf{\Lambda}' + \boldsymbol{\psi}$. The right-hand side is then

$$\begin{aligned} &\boldsymbol{\psi}^{-1}\mathbf{\Lambda}\mathbf{\Lambda}' + \mathbf{I} - \boldsymbol{\psi}^{-1}\mathbf{\Lambda}(\mathbf{I} + \mathbf{\Gamma})^{-1}\mathbf{\Gamma}\mathbf{\Lambda}' - \boldsymbol{\psi}^{-1}\mathbf{\Lambda}(\mathbf{I} + \mathbf{\Gamma})^{-1}\mathbf{\Lambda}' \\ &\qquad = \boldsymbol{\psi}^{-1}\mathbf{\Lambda}\mathbf{\Lambda}' + \mathbf{I} - \boldsymbol{\psi}^{-1}\mathbf{\Lambda}(\mathbf{I} + \mathbf{\Gamma})^{-1}(\mathbf{I} + \mathbf{\Gamma})\mathbf{\Lambda}' = \mathbf{I} \end{aligned}$$

as required. Substituting the expression for $\mathbf{\Sigma}^{-1}$ into (3.11) the left-hand side becomes

$$(\mathbf{\Lambda\Lambda}' + \boldsymbol{\psi} - \mathbf{S})\{\boldsymbol{\psi}^{-1} - \boldsymbol{\psi}^{-1}\mathbf{\Lambda}(\mathbf{I} + \mathbf{\Gamma})^{-1}\mathbf{\Lambda}\boldsymbol{\psi}^{-1}\}\mathbf{\Lambda}$$
$$= (\mathbf{\Lambda\Lambda}' + \boldsymbol{\psi} - \mathbf{S})\boldsymbol{\psi}^{-1}\mathbf{\Lambda}\{\mathbf{I} - (\mathbf{I} + \mathbf{\Gamma})^{-1}\mathbf{\Gamma}\}$$
$$= (\mathbf{\Lambda\Lambda}' + \boldsymbol{\psi} - \mathbf{S})\boldsymbol{\psi}^{-1}\mathbf{\Lambda}(\mathbf{I} + \mathbf{\Gamma})^{-1} \quad (3.13)$$

whence, setting the last expression equal to zero,

$$\mathbf{S}\boldsymbol{\psi}^{-1}\mathbf{\Lambda} = \mathbf{\Lambda}(\mathbf{I} + \mathbf{\Gamma}). \quad (3.14)$$

Pre-multiplying both sides by $\boldsymbol{\psi}^{-\frac{1}{2}}$ an alternative equation is

$$\boldsymbol{\psi}^{-\frac{1}{2}}\mathbf{S}\boldsymbol{\psi}^{-\frac{1}{2}}(\boldsymbol{\psi}^{-\frac{1}{2}}\mathbf{\Lambda}) = (\boldsymbol{\psi}^{-\frac{1}{2}}\mathbf{\Lambda})(\mathbf{I} + \mathbf{\Gamma}) \quad (3.15)$$

The second equation, (3.10), can likewise be expressed in terms of $\mathbf{\Lambda}$ and $\boldsymbol{\psi}$ as follows. Starting with $\mathbf{\Omega} = \mathbf{\Sigma}^{-1}(\mathbf{\Sigma} - \mathbf{S})\mathbf{\Sigma}^{-1}$ we substitute the expression for $\mathbf{\Sigma}^{-1}$ from (3.12) and use (3.13) to give

$$\mathbf{\Omega} = \boldsymbol{\psi}^{-1}(\mathbf{\Lambda\Lambda}' + \boldsymbol{\psi} - \mathbf{S})\boldsymbol{\psi}^{-1}.$$

If diag $\mathbf{\Omega}$ is to be zero then so must $\text{diag}(\mathbf{\Lambda\Lambda}' + \boldsymbol{\psi} - \mathbf{S})$ be zero, which implies that

$$\boldsymbol{\psi} = \text{diag}(\mathbf{\Lambda\Lambda}' - \mathbf{S}). \quad (3.16)$$

We have thus written the original pair of matrix equations in the alternative forms (3.15) and (3.16).

Suppose first of all that $\boldsymbol{\psi}$ is known; (3.15) may then be regarded as an equation in $\boldsymbol{\psi}^{-\frac{1}{2}}\mathbf{\Lambda}$. It will be satisfied if the following are true:

(a) $\boldsymbol{\psi}^{-\frac{1}{2}}\mathbf{\Lambda}$ is a matrix whose columns are any q eigenvectors of the matrix $\boldsymbol{\psi}^{-\frac{1}{2}}\mathbf{S}\boldsymbol{\psi}^{-\frac{1}{2}}$.
(b) $\mathbf{\Gamma}$ is a diagonal matrix such that $1 + \Gamma_{ii}$ is the eigenvalue associated with the eigenvector in the ith column of $\boldsymbol{\psi}^{-\frac{1}{2}}\mathbf{\Lambda}$.

The condition that $\mathbf{\Gamma}$ be diagonal emerges here purely as a mathematical requirement, but we note from (1.13) that it makes the conditional distribution of the y's independent. Later we shall see that it plays an important role in the alternative treatment given in Chapter 4.

To complete the solution for known $\boldsymbol{\psi}$ we have to find which q out of the p eigenvectors of $\mathbf{S}^* = \boldsymbol{\psi}^{-\frac{1}{2}}\mathbf{S}\boldsymbol{\psi}^{-\frac{1}{2}}$ are to be used in (a). To do this we express the log-likelihood in terms of the eigenvalues of the matrix $\mathbf{S}^*$. First we note that maximizing L of (3.6) is equivalent to maximizing

$$L' = \log|\mathbf{\Sigma}^{-1}\mathbf{S}| - \text{tr}\,\mathbf{\Sigma}^{-1}\mathbf{S}. \quad (3.17)$$

Using (3.12) and (3.13)

$$\mathbf{\Sigma}^{-1}\mathbf{S} = \boldsymbol{\psi}^{-1}\mathbf{S} - \boldsymbol{\psi}^{-1}\mathbf{\Lambda\Lambda}' = \boldsymbol{\psi}^{-\frac{1}{2}}(\mathbf{S}^* - \mathbf{\Lambda}^*\mathbf{\Lambda}^{*\prime})\boldsymbol{\psi}^{\frac{1}{2}}$$

where $\mathbf{\Lambda}^* = \boldsymbol{\psi}^{-\frac{1}{2}}\mathbf{\Lambda}$ and where all parameters take the values for which

L' is maximized. It follows that

$$\begin{aligned}\operatorname{tr}(\boldsymbol{\Sigma}^{-1}\mathbf{S}) &= \operatorname{tr}\{\psi^{-\frac{1}{2}}(\mathbf{S}^* - \boldsymbol{\Lambda}^*\boldsymbol{\Lambda}^{*\prime})\psi^{\frac{1}{2}}\} = \operatorname{tr}(\mathbf{S}^* - \boldsymbol{\Lambda}^*\boldsymbol{\Lambda}^{*\prime}) \\ &= \operatorname{tr}\mathbf{S}^* - \operatorname{tr}\boldsymbol{\Lambda}^*\boldsymbol{\Lambda}^{*\prime}. \end{aligned} \tag{3.18}$$

Now $\operatorname{tr}\mathbf{S}^* = \sum_{i=1}^{p} \theta_i$ where $\theta_1, \theta_2, \ldots, \theta_p$ are the eigenvalues of $\mathbf{S}^*$. The matrix $\boldsymbol{\Lambda}^*\boldsymbol{\Lambda}^{*\prime}$ has rank q because $\boldsymbol{\Lambda}^*$ is $p \times q$ and hence $p-q$ of its eigenvalues are zero. The remainder are also eigenvalues of $\mathbf{S}^* - \mathbf{I}$ since on replacing $\mathbf{S}^*$ in (3.15) by $\mathbf{I} + \boldsymbol{\Lambda}^*\boldsymbol{\Lambda}^{*\prime}$ we have

$$(\mathbf{I} + \boldsymbol{\Lambda}^*\boldsymbol{\Lambda}^{*\prime})\boldsymbol{\Lambda}^* = \boldsymbol{\Lambda}^* + \boldsymbol{\Lambda}^*\boldsymbol{\Gamma} = \boldsymbol{\Lambda}^*(\mathbf{I} + \boldsymbol{\Gamma}). \tag{3.19}$$

The non-zero eigenvalues of $\boldsymbol{\Lambda}^*\boldsymbol{\Lambda}^{*\prime}$ will thus be one less than the corresponding eigenvalues of $\mathbf{S}^*$. If we number the eigenvalues so that those which $\mathbf{S}^*$ and $\mathbf{I} + \boldsymbol{\Lambda}^*\boldsymbol{\Lambda}^{*\prime}$ have in common are $\theta_1, \theta_2, \ldots, \theta_q$ then

$$\operatorname{tr}\boldsymbol{\Lambda}^*\boldsymbol{\Lambda}^{*\prime} = \sum_{i=1}^{q} (\theta_i - 1)$$

and therefore

$$\operatorname{tr}(\boldsymbol{\Sigma}^{-1}\mathbf{S}) = \sum_{i=1}^{p} \theta_i - \sum_{i=1}^{q} (\theta_i - 1) = \sum_{i=q+1}^{p} \theta_i + q. \tag{3.20}$$

Next, we write

$$\boldsymbol{\Sigma} = \psi^{\frac{1}{2}}(\mathbf{I} + \boldsymbol{\Lambda}^*\boldsymbol{\Lambda}^{*\prime})\psi^{\frac{1}{2}}$$

whence

$$\begin{aligned}|\boldsymbol{\Sigma}| &= |\psi|\,|\boldsymbol{I} + \boldsymbol{\Lambda}^*\boldsymbol{\Lambda}^{*\prime}| \\ &= |\psi| \prod_{i=1}^{q} \theta_i\end{aligned}$$

since the determinant is the product of the eigenvalues found above. Similarly,

$$|\mathbf{S}| = |\psi^{\frac{1}{2}}\mathbf{S}^*\psi^{\frac{1}{2}}| = |\psi|\,|\mathbf{S}^*| = |\psi| \prod_{i=1}^{p} \theta_i$$

and hence

$$|\boldsymbol{\Sigma}^{-1}\mathbf{S}| = |\boldsymbol{\Sigma}|^{-1}\,|\mathbf{S}| = \prod_{i=q+1}^{p} \theta_i \tag{3.21}$$

and

$$L' = \sum_{i=q+1}^{p} (\log \theta_i - \theta_i) - q. \tag{3.22}$$

Since $\log x - x$ is a decreasing function of x for $x > 1$, L' will be a maximum if $\theta_{q+1}, \theta_{q+2}, \ldots, \theta_p$ are the smallest roots of $\mathbf{S}^*$. The eigenvectors chosen to provide the solution of (3.15) must therefore be those associated with the largest q roots as we might have expected.

We recall that the method set out above assumes that $\boldsymbol{\psi}$ is known. Later we shall see that there are approximate methods of estimating $\boldsymbol{\psi}$. The latter can also be used as starting values for an iterative solution of the full set of likelihood equations. One such method, called a zig-zag routine by Magnus and Neudecker (1986), and proposed by Rao (1955), was used in earlier versions of the SPSS computer package. Starting with an approximate estimate of $\boldsymbol{\psi}$, we obtain a first approximation to $\boldsymbol{\Lambda}$ by solving (3.15). This value of $\boldsymbol{\Lambda}$ can then be inserted into (3.16) to give an improved approximation to $\boldsymbol{\psi}$ and so on until convergence is obtained. Note that the diagonal elements of the fitted $\boldsymbol{\Sigma}$ will be precisely equal to the sample estimates in $\mathbf{S}$. A second approach replaces the use of (3.16) by a numerical optimization of the log-likelihood treated as a function of $\boldsymbol{\psi}$. This is due to Jöreskog (1963) who used the Fletcher–Powell method based on a second-degree approximation to L. Jennrich and Robinson (1969) proposed a Newton–Raphson method.

Goodness of Fit and Choice of q

If q is specified a priori, the goodness of fit of the factor model can be judged using the likelihood ratio statistic for testing the hypothesis $\boldsymbol{\Sigma} = \boldsymbol{\Lambda}\boldsymbol{\Lambda}' + \boldsymbol{\psi}$ (H_0) against the alternative that $\boldsymbol{\Sigma}$ is unconstrained (H_1). The statistic is then

$$\begin{aligned}-2\{L(H_0) - L(H_1)\} &= n\{\log|\hat{\boldsymbol{\Sigma}}| + \operatorname{tr}\hat{\boldsymbol{\Sigma}}^{-1}\mathbf{S} - \log|\mathbf{S}| - p\} \\ &= n\{\operatorname{tr}\hat{\boldsymbol{\Sigma}}^{-1}\mathbf{S} - \log|\hat{\boldsymbol{\Sigma}}^{-1}\mathbf{S}| - p\} \qquad (3.23)\end{aligned}$$

where $\hat{\boldsymbol{\Sigma}} = \hat{\boldsymbol{\Lambda}}\hat{\boldsymbol{\Lambda}}' + \hat{\boldsymbol{\psi}}$ is the estimated dispersion matrix. If $\boldsymbol{\psi} > \mathbf{0}$ this statistic is asymptotically distributed as χ^2 with degrees of freedom

$$\nu = \tfrac{1}{2}p(p+1) - \{pq + p - \tfrac{1}{2}q(q-1)\} = \tfrac{1}{2}\{(p-q)^2 - (p+q)\}. \qquad (3.24)$$

This is the difference between the number of parameters in $\boldsymbol{\Sigma}$ and the number of linear constraints imposed by the null hypothesis. Bartlett (1954) showed that the approximation can be improved by replacing n in (3.23) by $n - 1 - \tfrac{1}{6}(2p+5) - \tfrac{2}{3}q$. The behaviour of the test when n is small was investigated by Gweke and Singleton (1980) whose results suggest that it is adequate for n as low as 30 with one or two factors. However, Schönemann (1981) argues that this is too optimistic because they used untypical examples with high communalities. Since q is not usually specified in advance the test is often made the basis of a procedure for choosing the best value. Starting with $q = 1$ we then take successive values in turn until the fit of the model is judged to be adequate. Viewed as a testing procedure this is not strictly valid because it does not adjust the significance levels to allow for the sequential character of the test. It rather depends on regarding the p-value of the test as a measure of the adequacy of the model.

The trouble with a procedure of this kind is that the larger we make q the better the fit, but it provides us with no criterion for judging when to stop. An alternative approach is provided by Akaike's information criterion for model selection. The situation is that we have a set of linear models indexed by q and a selection has to be made. Akaike (1983) proposed the method for use in factor analysis and showed that it required q to be chosen to make

$$-2\{L(H_0) - L(H_1)\} - 2\nu$$

a minimum, where ν is the number of degrees of freedom given by (3.24). Note that $L(H_i)$ and ν are both functions of q. The criterion can be justified in a variety of ways but, in essence, it effects a trade-off between the bias introduced by fitting the wrong number of factors and the precision with which the parameters are estimated—as q is increased the bias decreases but the error increases. An alternative version due to Schwarz (1978) replaces the coefficient of ν by $\log_e n$. A further criterion based on the residuals of the fitted correlation matrix is proposed by Bozdogan and Ramirez (1986) who also report a comparison of all three criteria using Monte Carlo methods applied to models used in the study of Francis (1974). All methods perform reasonably well, but for the examples considered no method is uniformly best. These model selection ideas show considerable promise and their performance over a wide range of models merits further investigation.

Other criteria for selecting q which do not involve distributional assumptions will be mentioned in Section 3.3.

Sampling Variation of the Parameters

It is possible to obtain asymptotic variances and covariances for the estimators in the usual way by computing the second derivative matrix. Formulae are given by Lawley and Maxwell (1971, Chapter 5). Although it is often important to have standard errors of parameter estimates, the formulae given are rarely applicable in practice because they assume that the estimates have been obtained from the sample dispersion matrix rather than the correlation matrix. Lawley and Maxwell (1971) also give approximations applicable in this case, but they do not appear to have been incorporated into any statistical package. An alternative approach using bootstrap methods is given later. Swain (1975) shows that the same asymptotic properties hold if the estimates are obtained by minimizing any member of the class of functions $\sum_i f(\theta_i)$, where the θ's are the eigenvalues of $\mathbf{S}^*$.

The E-M *Algorithm*

An alternative approach to finding maximum likelihood estimators was suggested by Dempster, Rubin and Laird (1977) and has been examined further by Rubin and Thayer (1982, 1983) and Bentler and Tanaka (1983). We have discussed another application of this algorithm in Chapter 2 and will meet it again in Chapter 6. In the present case its use depends on the fact that if we knew the dispersion matrix of the y's we could estimate the parameters by regression methods. The method proceeds iteratively by choosing an arbitrary starting value for this matrix and then using it to estimate the parameters. These estimates enable us to obtain better estimates of the dispersion matrix and so on until convergence is reached. The method is rather slow and appears to offer no advantage over the well-tried methods already referred to.

Scale-invariant Estimation

In practice the factor model is almost always fitted using the sample correlation matrix rather than the dispersion matrix. The reason usually given for this is that the scaling adopted for the manifest variables is often arbitrary, especially in social science applications. Using correlations rather than covariances amounts to standardizing the x's using their sample standard deviations, and this ensures that changing the scale of the x's has no effect on the analysis. However, the joint distribution of the correlation coefficients is not the same as that of the covariances and so it is not obvious that the estimators obtained in the way described above will be true maximum likelihood estimators or that the asymptotic goodness of fit test will be valid. A good deal of confusion surrounds this topic so we shall approach the matter from first principles.

We start from the proposition that if the units of measurement of the x's are arbitrary then there can be no scientific interest in any aspect of the model which depends on the choice of units. The effect of a scale change on the model of (3.1) can be seen by transforming to $\mathbf{x}^* = \mathbf{C}\mathbf{x}$, where $\mathbf{C}$ is a diagonal matrix with positive elements. Expressed in terms of $\mathbf{x}^*$ the model becomes

$$\mathbf{x}^* = \mathbf{C}\boldsymbol{\mu} + \mathbf{C}\boldsymbol{\Lambda}\mathbf{y} + \mathbf{C}\mathbf{e} \tag{3.25}$$

with

$$D(\mathbf{x}^*) = \mathbf{C}\mathbf{\Lambda}\mathbf{\Lambda}'\mathbf{C} + \mathbf{C}\boldsymbol{\psi}\mathbf{C}.$$

In general, the parameters $\mathbf{\Lambda}$ and $\boldsymbol{\psi}$ do not therefore meet our requirements since their estimated values and the interpretation to be put upon them will depend on the scales of the x's. Suppose, however, that we choose $\mathbf{C} = (\operatorname{diag}\mathbf{\Sigma})^{-\frac{1}{2}}$ so that each x is divided by its theoretical standard deviation, then we may write (3.25) as

$$\mathbf{x}^* = \boldsymbol{\mu}^* + \mathbf{\Lambda}^*\mathbf{y} + \mathbf{e}^* \tag{3.26}$$

with

$$D(\mathbf{x}^*) = \mathbf{\Lambda}^*\mathbf{\Lambda}^{*\prime} + \boldsymbol{\psi}^* \quad \text{where} \quad \mathbf{\Lambda}^* = \mathbf{C}\mathbf{\Lambda} \quad \text{and} \quad \boldsymbol{\psi}^* = \mathbf{C}\boldsymbol{\psi}\mathbf{C}.$$

Clearly, any change in the scale of $\mathbf{x}$ now has no effect on the value of $\mathbf{x}^*$ and hence the parameter estimates obtained for $\mathbf{\Lambda}^*$ and $\boldsymbol{\psi}^*$ will be independent of the scaling of $\mathbf{x}$. These parameters will be said to be *scale-invariant* and it is on them that our interest will be centred. We shall also refer to $\mathbf{\Lambda}^*$ and $\boldsymbol{\psi}^*$ as the parameters of the model in *standard form.*

The important difference between this approach and that starting with the sample correlations is that in the latter case the x's would be standardized using the *sample* standard deviations. Their distribution would not then be normal, and that would be inconsistent with the assumptions we have made about the form and distribution of the right-hand side of (3.1).

To estimate $\mathbf{\Lambda}^*$ and $\boldsymbol{\psi}^*$ one could first estimate $\mathbf{\Lambda}$ and $\boldsymbol{\psi}$ for any arbitrary scaling of the x's using the sample dispersion matrix and then transform them by

$$\hat{\mathbf{\Lambda}}^* = (\operatorname{diag}\hat{\mathbf{\Sigma}})^{-\frac{1}{2}}\hat{\mathbf{\Lambda}} \quad \text{and} \quad \hat{\boldsymbol{\psi}}^* = (\operatorname{diag}\hat{\mathbf{\Sigma}})^{-\frac{1}{2}}\hat{\boldsymbol{\psi}}(\operatorname{diag}\hat{\mathbf{\Sigma}})^{-\frac{1}{2}}.$$

However it turns out that the usual practice of treating the correlation matrix as if it were the dispersion matrix does yield the maximum likelihood estimators of the scale-invariant parameters. This was shown by Krane and McDonald (1978) and an outline of the justification is as follows.

We re-parametrize the likelihood by writing $\mathbf{T} = \mathbf{C}^{-\frac{1}{2}}\mathbf{\Sigma}\mathbf{C}^{-\frac{1}{2}}$, where $\mathbf{C} = \operatorname{diag}\mathbf{\Sigma}$. From (3.6) we then have

$$\begin{aligned} -2L/n &= \text{Constant} + \log|\mathbf{C}^{\frac{1}{2}}\mathbf{T}\mathbf{C}^{\frac{1}{2}}| + \operatorname{tr}(\mathbf{C}^{\frac{1}{2}}\mathbf{T}\mathbf{C}^{\frac{1}{2}})^{-1}\mathbf{S} \\ &= \text{Constant} + \log|\mathbf{C}| + \log|\mathbf{T}| + \operatorname{tr}\mathbf{C}^{-\frac{1}{2}}\mathbf{T}^{-1}\mathbf{C}^{-\frac{1}{2}}\mathbf{S} \\ &= \text{Constant} + \log|\mathbf{C}| + \log|\mathbf{T}| + \operatorname{tr}\mathbf{T}^{-1}\mathbf{C}^{-\frac{1}{2}}\mathbf{S}\mathbf{C}^{-\frac{1}{2}}. \end{aligned} \tag{3.27}$$

The matrix $\mathbf{T}$ involves only the parameters $\mathbf{\Lambda}^*$, since its diagonal elements are now unities. The expression in (3.27) may be maximized in two stages, first with respect to **C** and then with respect to **T**. Krane and McDonald showed, as one might have anticipated, that $\mathbf{C} = \text{diag}\,\mathbf{S}$, so at the second stage the quantity to be maximized is

$$\log|\mathbf{T}| + \text{tr}\,\mathbf{T}^{-1}\mathbf{R}$$

where $\mathbf{R} = (\text{diag}\,\mathbf{S})^{-\frac{1}{2}}\mathbf{S}(\text{diag}\,\mathbf{S})^{-\frac{1}{2}}$ which is exactly what we would have done if we had treated the sample correlation matrix as a dispersion matrix. It is easily verified that the log-likelihood ratio statistic is the same whichever parametrization is used, so the procedure for estimation and testing goodness of fit is fully justified.

3.3 Fitting without normality assumptions

If nothing is assumed about the distributions of **y** and **e**, it remains true that the dispersion matrix predicted by the model is given by $\mathbf{\Sigma} = \mathbf{\Lambda\Lambda}' + \boldsymbol{\psi}$. We could then aim to estimate $\mathbf{\Lambda}$ and $\boldsymbol{\psi}$ in such a way that $\mathbf{\Sigma}$ was as close to **S**, the sample dispersion matrix, as is possible. For this we need some scalar measure of distance between $\mathbf{\Sigma}$ and **S** which must then be minimized with respect to the parameters. The distance function

$$\Delta(\mathbf{\Sigma}, \mathbf{S}) = -\text{tr}\,\mathbf{\Sigma}^{-1}\mathbf{S} + \log|\mathbf{\Sigma}^{-1}\mathbf{S}| \tag{3.28}$$

which arose in the last section is one possibility. It will only yield maximum likelihood estimators if the normal assumptions hold, but if (3.28) were regarded as a reasonable way of measuring distance the estimators would be justified on a much broader basis.

There are many other possible measures of distance which could be used in place of (3.28). It is natural to turn to least squares ideas. A simple unweighted least squares criterion would be

$$\begin{aligned}\Delta_1 &= \sum_{i=1}^{p}\sum_{h=1}^{p}(s_{ih} - \sigma_{ih})^2 \\ &= \text{tr}(\mathbf{S} - \mathbf{\Sigma})^2.\end{aligned} \tag{3.29}$$

Another, suggested by the role played by the matrix $\mathbf{S}^* = \boldsymbol{\psi}^{-\frac{1}{2}}\mathbf{S}\boldsymbol{\psi}^{-\frac{1}{2}}$ in maximum likelihood estimation, is

$$\begin{aligned}\Delta_2 &= \text{tr}\{(\mathbf{S}^* - \mathbf{\Sigma}^*)^2\} \\ &= \text{tr}\{\boldsymbol{\psi}^{-1}(\mathbf{S} - \mathbf{\Sigma})^2\boldsymbol{\psi}^{-1}\} = \text{tr}\{(\mathbf{S} - \mathbf{\Sigma})\boldsymbol{\psi}^{-1}\}^2.\end{aligned} \tag{3.30}$$

These are both special cases of a general class of measures stemming from fundamental work by Browne (1982, 1984) (see also Tanaka and Huba (1985)) which may be written

$$\Delta = \text{tr}\{(\mathbf{S} - \mathbf{\Sigma})\mathbf{V}\}^2. \tag{3.31}$$

These arise from a generalized least squares approach which allows the deviations $\{s_{ij} - \sigma_{ij}\}$ to be weighted in various ways. We shall not develop these ideas here, but reference to the authors mentioned above will show that these methods lead to robust methods of estimation under a range of distributional assumptions. The case $\mathbf{V} = \mathbf{S}^{-1}$ is also discussed in Anderson (1984, Section 14.3.4). The attraction of Δ_1 and Δ_2 is that the optimization requires the solution of a simple eigenvalue problem. In the case of Δ_1 the function to be minimized may be written

$$\Delta_1 = \sum_{i=1}^{p} \sum_{h=1}^{p} \left(s_{ih} - \delta_{ih}\psi_i - \sum_{j=1}^{q} \lambda_{ij}\lambda_{hj} \right)^2.$$

Differentiating with respect to λ_{rs}

$$\frac{\partial \Delta_1}{\partial \lambda_{rs}} = 4\left\{ - \sum_{i=1}^{p} (s_{ir} - \delta_{ir}\psi_i)\lambda_{is} + \sum_{i=1}^{p} \lambda_{is} \sum_{j=1}^{q} \lambda_{ij}\lambda_{rj} \right\}$$

$$(r = 1, 2, \ldots, p; s = 1, 2, \ldots, q)$$

or

$$\frac{\partial \Delta_1}{\partial \mathbf{\Lambda}} = 4\{\mathbf{\Lambda}(\mathbf{\Lambda}'\mathbf{\Lambda}) - (\mathbf{S} - \boldsymbol{\psi})\mathbf{\Lambda}\}$$

which gives the estimating equations

$$(\mathbf{S} - \boldsymbol{\psi})\mathbf{\Lambda} = \mathbf{\Lambda}(\mathbf{\Lambda}'\mathbf{\Lambda}). \tag{3.32}$$

Differentiating with respect to ψ_r,

$$\frac{\partial \Delta_1}{\partial \psi_r} = -2\left(s_{rr} - \psi_r - \sum_{j=1}^{q} \lambda_{rj}^2 \right)$$

or

$$\operatorname{diag} \frac{\partial \Delta_1}{\partial \boldsymbol{\psi}} = -\operatorname{diag} \mathbf{S} + \boldsymbol{\psi} + \operatorname{diag} \mathbf{\Lambda}\mathbf{\Lambda}'$$

leading to

$$\boldsymbol{\psi} = \operatorname{diag}(\mathbf{S} - \mathbf{\Lambda}\mathbf{\Lambda}'). \tag{3.33}$$

These estimating equations should be compared with those for maximum likelihood in (3.15) and (3.16) when it will be seen that a similar method of solution will apply. Suppose first that $\boldsymbol{\psi}$ is known then (3.32) will be satisfied if:

(a) the columns of $\mathbf{\Lambda}$ consist of any q eigenvectors of $\mathbf{S} - \boldsymbol{\psi}$
(b) $\mathbf{\Lambda}\mathbf{\Lambda}'$ is a diagonal matrix with elements equal to the eigenvalues of $\mathbf{S} - \boldsymbol{\psi}$ associated with the vectors in $\mathbf{\Lambda}$.

Thus if we have a starting value for $\boldsymbol{\psi}$, the solution of (3.32) will yield a first approximation to $\boldsymbol{\Lambda}$ which can then be inserted in (3.33) to give a second estimate of $\boldsymbol{\psi}$, and then the cycle can be continued until convergence occurs. The question of which eigenvectors of $\mathbf{S}-\boldsymbol{\psi}$ are to be included in $\boldsymbol{\Lambda}$ can be answered as follows.

$$\begin{aligned}\Delta_1 &= \operatorname{tr}(\mathbf{S}-\boldsymbol{\psi}-\boldsymbol{\Lambda}\boldsymbol{\Lambda}')^2 = \operatorname{tr}(\mathbf{S}-\boldsymbol{\psi})^2 - 2\operatorname{tr}(\mathbf{S}-\boldsymbol{\psi})\boldsymbol{\Lambda}\boldsymbol{\Lambda}' + \operatorname{tr}(\boldsymbol{\Lambda}\boldsymbol{\Lambda}')^2 \\ &= \operatorname{tr}(\mathbf{S}-\boldsymbol{\psi})^2 - \operatorname{tr}(\boldsymbol{\Lambda}\boldsymbol{\Lambda}')^2, \quad \text{using} \quad (3.32).\end{aligned}$$

Now $\boldsymbol{\Lambda}\boldsymbol{\Lambda}'$ has $(p-q)$ zero eigenvalues because it is of rank q. The remainder are also eigenvalues of $\mathbf{S}-\boldsymbol{\psi}$ since if we replace $\mathbf{S}-\boldsymbol{\psi}$ in (3.32) by $\boldsymbol{\Lambda}\boldsymbol{\Lambda}'$ the equation is obviously satisfied. Let the eigenvalues which the two matrices have in common be $\theta_1, \theta_2, \ldots, \theta_q$ and the remaining eigenvalues of $\mathbf{S}-\boldsymbol{\psi}$ be $\theta_{q+1}, \ldots, \theta_p$ then

$$\Delta_1 = \sum_{i=1}^{p} \theta_i^2 - \sum_{i=1}^{q} \theta_i^2 = \sum_{i=q+1}^{p} \theta_i^2. \tag{3.34}$$

For this to be a minimum, $\theta_{q+1}, \ldots, \theta_p$ must be the smallest eigenvalues and hence $\boldsymbol{\Lambda}$ must consist of the eigenvectors associated with the q largest eigenvalues.

This method is known as the *principal factor* (*or axis*) *method* because of its similarity to principal components analysis to which it is equivalent if $\boldsymbol{\psi}=\mathbf{0}$. If we use the correlation matrix $\mathbf{R}$ instead of $\mathbf{S}$ the estimates obtained for the scale-invariant parameters $\boldsymbol{\Lambda}^*$ and $\boldsymbol{\psi}^*$ will not be identical to those arrived at by first using $\mathbf{S}$ and then transforming as they were with the maximum likelihood method.

Estimation for Δ_2 using (3.30) proceeds in an exactly similar manner. In fact, since $\boldsymbol{\Sigma}^* = \boldsymbol{\Lambda}^*\boldsymbol{\Lambda}^{*\prime} + \mathbf{I} = (\boldsymbol{\psi}^{-\frac{1}{2}}\boldsymbol{\Lambda})(\boldsymbol{\psi}^{-\frac{1}{2}}\boldsymbol{\Lambda})' + \mathbf{I}$ all we have to do is to replace $\boldsymbol{\Lambda}$ by $\boldsymbol{\Lambda}^* = \boldsymbol{\psi}^{-\frac{1}{2}}\boldsymbol{\Lambda}$ in (3.32) and $\mathbf{S}-\boldsymbol{\psi}$ by $\mathbf{S}^*-\mathbf{I}$. The estimating equation for $\boldsymbol{\Lambda}^*$ is then identical with that for maximum likelihood given in (3.15). Rather surprisingly it turns out that the differing distance functions both lead to the eigenvalues and vectors of $\mathbf{S}^*$. However, this is true only for fixed $\boldsymbol{\psi}$. If we bring $\boldsymbol{\psi}$ into the picture its partial derivatives will be different from those in the maximum likelihood case and, in fact, a good deal more complicated.

For Δ given by (3.31) with $\mathbf{V}=\mathbf{S}^{-1}$ the same equation for $\boldsymbol{\Lambda}$ for fixed $\boldsymbol{\psi}$ is obtained, but the (implicit) estimating equation for $\boldsymbol{\psi}$ is

$$\operatorname{diag} \mathbf{S}^{-1}\{(\boldsymbol{\Lambda}\boldsymbol{\Lambda}'+\boldsymbol{\psi})-\mathbf{S}\}\mathbf{S}^{-1} = \operatorname{diag} \mathbf{0};$$

see Anderson (1984).

We have noted that if $\boldsymbol{\psi}$ were known, these methods would be much simpler, being straightforward eigen problems. Since $\boldsymbol{\psi}$ only enters into the diagonal elements of $\boldsymbol{\Sigma}$, there is a prospect of avoiding the difficulties by eliminating these terms from Δ_1. We would then minimize

$$\Delta_1' = \sum_{\substack{i=1 \\ i \neq h}}^{p} \sum_{h=1}^{p} \left(s_{ih} - \sum_{j=1}^{q} \lambda_{ij} \lambda_{hj} \right)^2. \tag{3.35}$$

This approach is described in Harman and Jones (1966) and is usually known as the "minres" method. Various methods of obtaining estimates have been given by Comrey (1962), Comrey and Ahumada (1964), Okamoto and Ihara (1983) and Zegers and ten Berge (1983) (see also Chapter 6). Unfortunately the omission of the diagonal terms destroys the structure which led to the easily solved eigen equations. Given that the minres method is unweighted and that the iterative methods for the Δ-family are well within the scope of modern computers, the method offers few advantages and is not widely used. There are, however, problems in the factor analysis of categorical data, discussed later, where the other methods are not available and where the minres method is useful.

Consistency

In all the foregoing methods we have tacitly assumed that the parameters can be consistently estimated. A necessary condition for this to be possible is that there shall be at least as many sample statistics as there are parameters to be estimated. The number of parameters in the q-factor model is $pq + p$, but in order to obtain a unique solution we have to impose $\frac{1}{2}q(q-1)$ constraints (usually by requiring the off-diagonal elements of $\mathbf{\Lambda}'\boldsymbol{\psi}\mathbf{\Lambda}$ to be zero). The number of free parameters is then

$$pq + p - \tfrac{1}{2}q(q-1).$$

The sample dispersion matrix $\mathbf{S}$ had $\frac{1}{2}p(p+1)$ distinct elements, so for consistent estimation to be possible we must have

$$\tfrac{1}{2}p(p+1) - pq - p + \tfrac{1}{2}q(q-1) = \tfrac{1}{2}[(p-q)^2 + (p+q)] \geqslant 0. \tag{3.36}$$

Equation (3.36) implies that there is an upper bound to the number of factors which can be fitted given by

$$q \leqslant \{2p + 1 - (8p+1)^{\frac{1}{2}}\}. \tag{3.37}$$

The condition (3.36) is not sufficient because it does not guarantee that the estimates of the ψ_i's will be non-negative. Building on earlier work by Anderson and Rubin (1956), Kano (1983, 1986a, 1986b) has provided general conditions under which maximum likelihood and generalized least squares estimators are consistent. However, in Kano (1986a) he provides an example of a model which does not admit any consistent estimator. The question of consistency when p is not fixed has been investigated in Kano (1986b).

Approximate Methods for Estimating $\boldsymbol{\psi}$

The iterative methods require a starting value for $\boldsymbol{\psi}$, and although one can use an arbitrary value such as $\boldsymbol{\psi} = \operatorname{diag} \mathbf{S}$ there are approximations which will reduce the number of iterations required. Apart from their practical value they give some insight into the interpretation of the analysis. Returning to (3.12), we have

$$\operatorname{diag} \boldsymbol{\Sigma}^{-1} = \boldsymbol{\psi}^{-1} - \operatorname{diag}\{\boldsymbol{\psi}^{-1}\boldsymbol{\Lambda}(\mathbf{I}+\boldsymbol{\Gamma})^{-1}\boldsymbol{\Lambda}'\boldsymbol{\psi}^{-1}\}. \tag{3.38}$$

Multiplying out the last term and remembering that $\boldsymbol{\Gamma}$ is diagonal, we find

$$\sigma^{ii} = \psi_i^{-1}\left[1 - \sum_{j=1}^{q} \bigg/ \left(1 + \sum_{r=1}^{p} \frac{\lambda_{rj}^2}{\psi_r}\right)\right], \quad (i = 1, 2, \ldots, p) \tag{3.39}$$

from which it follows that

$$\sigma^{ii} \leqslant \psi_r^{-1}. \tag{3.40}$$

Since σ^{ii} can be estimated from the inverse of $\mathbf{S}$, we can estimate an upper bound for ψ_i and this may be used as a starting value for the iteration. Noting that the last expression in the bracket of (3.39) is a term of order q divided by a term of order p, we may expect the bound to be a good approximation if p is large relative to q. Increasing p for fixed q will improve the approximation and, roughly speaking, as p increases without limit the bound is attained.

This result may be given another interpretation: s^{ii} which we would use as an approximation may be expressed using standard regression results as

$$s^{ii} = s_{ii}(1 - R_i^2) \tag{3.41}$$

where R_i^2 is the multiple correlation coefficient of x_i regressed on the remaining x's. Now

$$s_{ii}(1 - R_i^2) = \operatorname{var}(x_i \mid x_1, x_2, \ldots, x_{i-1}, x_{i+1}, \ldots, x_p)$$

and

$$\psi_i = \operatorname{var}(x_i \mid \mathbf{y}).$$

We would expect the latter to be smaller than the former because $\mathbf{y}$ is not precisely determined by any finite set of x's (see (1.13)). A simpler but less precise bound for ψ_i follows from the fact that $R_i^2 \geqslant \max_{j \neq i} r_{ij}^2$, hence an estimated upper bound for ψ_i is $s_{ii}(1 - \max r_{ij}^2)$.

Goodness of Fit and Choice of q

Little appears to be known about the sampling behaviour of the methods of fitting discussed in this section, but there is an important result due to Amemiya and Anderson (1985) which shows that the limiting χ^2 distribution of the goodness of fit statistic of (3.23) is valid under very general circumstances. They show that if the elements of $\mathbf{e}$ are independent and if $\mathbf{y}$ and $\mathbf{e}$ have finite second moments then (3.23) calculated using the maximum likelihood estimators has the same distribution as in the normal case. This result also holds for another goodness of fit statistic which is sometimes used, namely

$$\tfrac{1}{2}n \operatorname{tr}\{(\mathbf{S} - \hat{\mathbf{\Sigma}})\hat{\mathbf{\Sigma}}^{-1}\}^2$$

which, in the normal case, is asymptotically equivalent to (3.23). This is a further reason for using the maximum likelihood method of fitting, whether or not one makes the normality assumptions.

One consequence of these results is that we can use the same methods for choosing q as were proposed for the maximum likelihood method in the last section. There are also two other methods which do not depend on distributional assumptions. Both are based on the eigenvalues of the sample correlation matrix and the role which they have in principal components analysis. The Kaiser–Guttman criterion chooses q equal to the number of eigenvalues greater than one. The rationale is that the average contribution of a manifest variable to the total variation is one, and that a principal component which did not contribute at least as much variation as a single variable represents no advantage. The carry-over of this argument from principal components to factor analysis rests on the similarity between the two techniques noted in Chapter 1. Simulation results obtained by Fachel (1986) suggest that if p is large this method is likely to overestimate q.

The second method due to Cattell is known as the "scree test". If the eigenvalues are plotted against their rank order they will lie on a decreasing curve. One then looks for an "elbow" in the curve, as this would indicate the point at which the further addition of factors shows diminishing returns in terms of variation explained. A simulation study by Hakstian, Rogers and Cattell (1982) comparing these two methods with the use of the likelihood ratio statistic does not lead to clear-cut conclusions.

Sampling Variation of the Parameters

Similarly there is a lack of theory about the standard errors of the parameter estimates, though here again one would expect the asymptotic theory for the normal case to be applicable. Nevertheless it is possible to obtain some information by simulation and bootstrap or jackknife techniques, and although this is relatively underdeveloped territory the results obtained to date suggest a degree of caution in using factor analysis that is seldom found in published studies. Tucker, Koopman and Linn (1969) carried out a detailed simulation study, and Seber (1984) reported a series of simulation studies on samples of size 50 by Francis (1974). Pennell (1972) demonstrated how the jackknife technique can be used to find confidence intervals for factor loadings. Here we describe an application of the bootstrap technique due to Chatterjee (1984). The bootstrap technique is due to Efron and described by him in Efron (1982) and also in Efron and Tibshirani (1986). If we have a sample of size n we draw repeated samples of size n *with replacement* from the original sample. We shall then have a sequence of samples each having multiple occurrences of some of the sample individuals. The factor model would then be fitted to each sample so drawn. The basic result of bootstrap theory is that the empirical distributions of the parameter estimates obtained by this method are asymptotically the same as the sampling distribution of those parameters in sampling from the population from which the original sample was drawn. The method has been applied to a number of other statistical techniques, but this appears to be the only investigation of factor analysis by bootstrap methods.

Chatterjee used data taken from Johnson and Wichern (1982) concerning seven variables measured on 50 randomly chosen salesmen. Three variables were measures of sales performance and four were from tests of aptitude. The method of fitting was the principal factor method with the ψ's set equal to zero (i.e. a principal components analysis). One would have preferred a method which allowed estimation of the ψ's, but with only two or three factors turning out to be significant the results are not likely to be seriously affected and the computational ease of this method commended it for this particular study. The question of how many repeated samples to draw was settled empirically and it appeared that 300 gave reasonable stability. Some of the results are given in Table 3.1 for factor loadings (standardized λ's) for the first three factors. The table shows that the first factor is well determined but that we should be wary of attributing much significance to the other two. None of the loadings for factor 3 differ from zero by much more than their standard deviation and only two or three of those for factor 2 come anywhere near significance. An advantage of the bootstrap method is that we can also look at the frequency distributions of the estimators. Chatterjee (1984) gives a number of examples of which that given in Table 3.2 is particularly instructive. We notice that the large standard deviation is due to the extreme skewness of the distribution. Six percent of the samples

Table 3.1 Estimated loadings and their standard deviations for Johnson and Wichern's data

Variable	Original estimates Factor 1	Factor 2	Factor 3	Bootstrap estimates on 300 samples Factor 1	S.D.	Factor 2	S.D.	Factor 3	S.D.
1	.973	−.110	.054	.972	.006	−.096	.055	.038	.052
2	.943	.029	.312	.945	.013	.010	.091	.192	.158
3	.945	.010	−.144	.943	.020	.014	.079	.076	.168
4	.660	.646	−.318	.657	.095	.577	.276	−.179	.254
5	.783	.286	−.005	.774	.064	.268	.165	.041	.373
6	.649	−.620	−.427	.644	.093	−.516	.358	−.266	.229
7	.914	−.193	.306	.916	.017	−.185	.091	.197	.173

Table 3.2 Frequency distribution of factor 2 loading for variable 4 (300 samples)

Midpoint of interval	.9	.8	.7	.6	.5	.4	.3	.2	.1	0	−.1	−.2	−.3	−.4	−.5	−.6	−.7
Frequency	7	43	111	52	40	22	4	2	1	0	0	3	1	6	2	5	1

actually give negative loadings (if *all* loadings on a factor were negative the signs should have been reversed) but even on the positive half of the scale the scatter is considerable.

Although this example is very limited in both scope and method it provides a warning against taking estimated loadings at their face value for sample sizes as small as 50. The results reported by Seber (1984) and others support this. Much more work is urgently needed to extend and consolidate our limited knowledge in this important area. Results reported in later chapters for categorical variables strongly suggest that much larger samples (say 500 or more) are needed if parameters are to be estimated with precision. The lack of reproducibility of the results of factor analysis which has somewhat tarnished its image among practitioners doubtless owes much to the use of inadequate sample sizes. The bootstrap technique is relatively expensive in computer time and could not be used on a routine basis but it could be used on a range of typical cases as a means of accumulating experience on sampling behaviour.

A much simpler approach which can always be used is what is known as cross-validation. By splitting the sample randomly into two (or more) equal parts and fitting the model to each part, some, limited, idea can be gained about the stability of the estimates. An illustration for the latent class model, due to Pickering and Forbes (1984), was reported in Chapter 2.

3.4 Interpretation

After the model has been fitted the analyst will usually wish to "interpret" the parameters and the factors; that is to identify in the latter case what it is that the dimensions of the factor space represent. The manner in which this is done will depend very much on the disciplinary context in which the study is carried out. An educational tester, for example, may have deliberately chosen the x's as indicators of some well-defined ability and the same is often true in psychology more generally. In other areas of social research the analysis may be exploratory with no pre-conceptions about the underlying dimensions. In the first case the aim is primarily to confirm an existing theory and in the second to aid in the formation of relevant concepts for a theory. These matters are philosophical rather than statistical, but since interpretation is often facilitated by further analysis they do raise methodological issues. In particular we have already noted that transformation of the factor space leaves the fit of the model unchanged, and we therefore need ways of exploring the set of possible transformations.

Here we shall describe the usual methods of interpreting the parameters. An alternative approach is given in Chapter 4. We start from the fact that

$$\operatorname{var}(x_i) = \lambda_{i1}^2 + \lambda_{i2}^2 + \ldots + \lambda_{iq}^2 + \psi_i, \quad (i = 1, 2, \ldots, p). \quad (3.42)$$

This enables us to interpret the parameters in terms of the proportion of the variance of x_i which is accounted for by each of the latent variables on which it depends. Thus

$$\psi_i \Big/ \Big(\sum_j \lambda_{ij}^2 + \psi_i\Big) = \psi_i^* \quad \text{(the scale-invariant parameter)}$$

is the proportion of variance accounted for by the error term; its complement is therefore the proportion due to all the factors and is known as the *communality*. Similarly

$$\lambda_{ij}^2 \Big/ \Big(\sum_j \lambda_{ij}^2 + \psi_i\Big) = \lambda_{ij}^{*2}$$

is the proportion of x_i's variance attributable to y_i. If the model has been fitted using the correlation matrix we shall have estimated the scale-invariant parameters directly. Since this will usually be the case, and for typographical convenience, we shall omit the asterisks and assume

$$\sum_{j=1}^{q} \lambda_{ij}^2 + \psi_i = 1 \quad (3.43)$$

for the remainder of this section.

It is also useful to consider the decomposition of the total variance of $\mathbf{x}$ by summing both sides of (3.38) over i. We then have

$$\sum_{i=1}^{p} \text{var}(x_i) = \sum_{i=1}^{p} \lambda_{i1}^2 + \sum_{i=1}^{p} \lambda_{i2}^2 + \ldots + \sum_{i=1}^{p} \lambda_{iq}^2 + \sum_{i=1}^{p} \psi_i. \qquad (3.44)$$

$\sum_{i=1}^{p} \lambda_{ij}^2$ is then the proportion of the total variance of $\mathbf{x}$ which is explained by y_j; it thus provides a measure of the importance of y_j in determining the variability of $\mathbf{x}$.

In order to interpret the factors, as distinct from the parameters, we note that λ_{ij} is the standardized regression coefficient of y_j (sometimes called a beta coefficient). It measures the strength of the dependence of x_i on y_j and is, in fact, also the correlation coefficient of x_i and y_j. In interpreting factor y_j one would look at the relative values of $\lambda_{1j}, \lambda_{2j}, \ldots, \lambda_{pj}$ and at their signs. Manifest variables for which λ_{ij} is large are thus more closely linked with the factor than those for which it is not. If for example all λ's have the same sign and are of comparable magnitude, one would seek to interpret the factor in question in terms of what the set of manifest variables has in common. Such a factor is often called a general factor. Another common situation arises when the λ's fall into two groups, one positive and the other negative and all of similar absolute magnitude. Here one would ask what one group has in common which the other lacks and vice versa. Such a factor is known as a bi-polar factor.

Another set of factor loadings (λ's) which is relatively easy to interpret is one where many of the λ's are close to zero and the remainder are of the same sign and similar magnitude. We interpret this in exactly the same way as a general factor except that we consider the sub-set of manifest variables with substantial loadings rather than the whole set. In some applications we may have prior knowledge about which groupings of manifest variables are linked to particular factors and we shall consider how this information might be used below.

If the pattern of loadings revealed does not lend itself to interpretation in this way it may be possible to transform the factor space so that such an interpretation is possible. This may be achieved by rotation.

Orthogonal Rotation

We already know that we are at liberty to make any linear transformation of the factor space without affecting the fit of the model. The loadings for the transformed space are found by applying the inverse transformation to the loadings matrix $\mathbf{\Lambda}$ as shown in (3.3).

This can easily be visualized graphically for the case of two factors. If we plot the pairs $(\lambda_{i1}, \lambda_{i2})$ for all i, we may be able to see what transformation of $\mathbf{\Lambda}$ might produce a more easily interpretable pattern. This is illustrated in the example which follows.

If there are more than two factors the search for a suitable rotation is much more difficult. It is then desirable to have an algorithm which can be programmed to search for a transformation which has the desired properties. The aim of making the loadings on a particular factor either small or large can be formalized by requiring that the variance of $(\lambda_{1j}^2, \lambda_{2j}^2, \ldots, \lambda_{pj}^2)$ be as large as possible. We take the squares because this ensures that each value is between zero and one, and the maximization will then force the λ's to either end of the range. A rotation found by this method is called a *varimax* rotation and was proposed by Kaiser (1958). In a modified version λ_{ij}^2 is replaced by $\lambda_{ij}^2/(1-\psi_i)$, this representing the proportion of the common variance of x_i that is attributable to y_j. The maximization is achieved iteratively.The details are given in Magnus and Neudecker (1986), Lawley and Maxwell (1971) and, for $q = 2$, in Mardia *et al.* (1979). An orthogonal rotation will not necessarily provide a more easily interpretable solution, particularly if there is a dominant general factor.

A rather special orthogonal transformation is when the rotation is through two right angles which simply changes the signs of all the loadings on the factor. This essentially leaves the factors unaltered because changing the sign of the loadings is the same as changing the sign, and hence the direction of measurement, of the factor.

Oblique Rotation

If an orthogonal rotation does not produce an interpretable pattern of loadings it may be possible to do so by admitting non-orthogonal (oblique) transformations. The loss of orthogonality complicates the interpretation of the parameters and the factors, but there are often good substantive reasons for preferring such a solution, as our example will show. Suppose, for example, that one is seeking to isolate two factors called verbal ability and arithmetical ability in an educational test. There may be reason to suppose that these two factors are correlated and, if so, we shall not uncover them if we are limited in our search to orthogonal factors. We can, therefore, generalize the varimax idea to include oblique axes while still looking for factors which load on relatively few manifest variables.

A $\mathbf{\Lambda}$-matrix which has a pattern of zeros (or near-zeros) such that each factor loads on distinct non-overlapping sub-sets of the manifest variables is said to possess *simple structure*. This term was coined by Thurstone in the context of the factor analysis of human abilities where it had a particular connotation. However, it is useful to use it more widely because any such matrix is often relatively easy to interpret.

The theory of the Promax method is given in Lawley and Maxwell (1971, Section 6.4) and other algorithms are available in the various statistical packages (e.g. Oblimin in SPSS). Details of what is currently available will be found in Tabachnick and Fidell (1983).

If the factors are correlated the question of interpretation must be reconsidered. The simple decomposition of (3.42) is no longer possible because there will be additional terms involving correlations between the y's; ψ_i may, however, still be interpreted in the same way as before. The λ's can no longer be interpreted as correlation coefficients but they are still standardized regression coefficients indicating the contribution of each factor to each manifest variable. One can therefore still interpret factors by reference to the magnitude and pattern of the elements of $\mathbf{\Lambda}$.

An interesting extension of the analysis of results obtained from an oblique rotation in a study of motivation in smoking is given in Russell, Peto and Patel (1974). Having obtained an interpretable set of oblique factors they go on to factor-analyse the matrix of correlations between the factors in an attempt to isolate even more fundamental dimensions. This practice also has its roots in the debates in the first half of the century about the nature of human abilities. Following Spearman's theory of a single factor model of intelligence and Thurstone's alternative theory of a cluster of correlated specific abilities, Burt attempted to reinstate the idea of an underlying "g" factor of general ability by analysing the correlation matrix of abilities. This raises debatable issues in the philosophy of science, but from a statistical point of view such analysis has no new empirical content. It rather sacrifices some of the results of the first level analysis in return for greater simplicity in the interpretation .

Example

Many of the foregoing ideas will now be illustrated using data from Smith and Stanley (1983) on ability scores. The full study was concerned with the relationship between reaction times and intelligence tests scores. Here we consider only the factor analysis of the ability variables. Scores were available for 112 individuals on the following six variables:

(1) A non-verbal measure of general intelligence (Spearman's g) using Cattell's culture-fair test.
(2) Picture completion test

(3) Block design
(4) Mazes
(5) Reading comprehension
(6) Vocabulary

Full details may be found in the original paper. The correlation coefficients, covariances and variances, supplied by the authors, are set out in Table 3.3. The data were first analysed using the maximum

Table 3.3 Correlation coefficients (right upper) and variances and covariances (left lower) for Smith and Stanley's data

	1	2	3	4	5	6
1	24.641	.466	.552	.340	.576	.510
2	5.991	6.700	.572	.193	.263	.239
3	33.520	18.137	149.831	.445	.354	.356
4	6.023	1.782	19.424	12.711	.184	.219
5	20.755	4.936	31.430	4.757	52.604	.794
6	29.701	7.204	50.753	9.075	66.762	135.292

likelihood routine in the SPSS-X package with variances and covariances as input. The choice of the number of factors to fit is limited to one or two because with $q \geqslant 3$ inequality (3.37) is violated. The values of the likelihood ratio statistic are:

1-factor: $\chi^2 = 75.56$ with 9 degrees of freedom

2-factor: $\chi^2 = 6.07$ with 4 degrees of freedom.

Whether we use the p-values or Akaike's criterion it is clear that the evidence strongly favours a 2-factor model. The correlation matrix has two eigenvalues greater than one (3.08 and 1.14) and the first two principal components together account for 70% of the variation in x. These facts also point to a two-factor model.

Since the x's have no natural common scale it is sensible to base the interpretation on the standardized loadings and communalities. This may be done by transforming the output values to λ^*'s and ψ^*'s as described in Section 3.4. For maximum likelihood estimation this will give the same results as if the correlation matrix had been analysed, but for the principal axis method (PAF) the results will be slightly different for the reasons given by Krane and McDonald (1978) noted earlier. For purposes of comparison the correlation matrix was used for the PAF method, and the results using the SPSS-X package are given in Table 3.4. The communality estimates are similar for both methods but the factor loadings are not close. However, this discrepancy is more apparent than real since, as Fig. 3.1 shows, the one is approximately an orthogonal rotation of the other. The difference

Table 3.4 Factor loadings and communalities estimated by two methods for Smith and Stanley's data

	Maximum likelihood			Principal axis (least squares)		
i	λ_{i1}	λ_{i2}	ψ_i	λ_{i1}	λ_{i2}	ψ_i
1	.64	.37	.54	.75	.07	.57
2	.34	.54	.41	.52	.32	.38
3	.46	.76	.78	.75	.52	.84
4	.25	.41	.23	.39	.22	.20
5	.97	−.12	.95	.82	−.51	.93
6	.82	−.03	.67	.73	−.38	.68

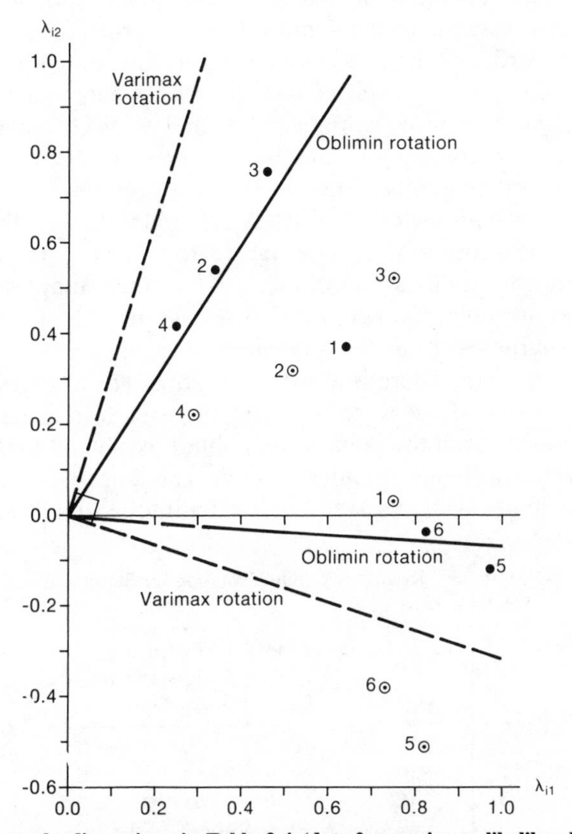

Fig. 3.1 Factor loadings given in Table 3.4 (dots for maximum likelihood, circles for principal axis). The axes for the Varimax and Oblimin rotations are shown for the maximum likelihood estimators only.

arises because of the different constraints imposed by the two methods to make the solution unique.

In neither case do the loadings correspond closely to any of the easily interpretable patterns discussed earlier. It is true that the first factor in each case has positive and reasonably large loadings and so might be regarded as a general factor, but the second factor then has no simple interpretation. A varimax rotation was carried out and the new axes are shown on Fig. 3.1 for the maximum likelihood solution. Referred to the new axes all loadings will be positive, but otherwise the position is not materially altered. Much greater insight comes when we consider oblique rotations. Fig. 3.1 also shows that oblique axes can be found such that all points but one lie very close to them.

The rotated loadings obtained by the Oblimin method of SPSS-X are given in Table 3.5 which shows something close to simple structure. The fact that the maximum likelihood and principal axis methods give essentially the same solution is partly obscured by the fact that the order of the two factors is reversed and that in the case of one factor there is a reversal of sign. It now appears that there is one factor loading heavily on variables 2, 3 and 4 which has to do with problem-solving ability, and another loading on variables 5 and 6 relating to verbal ability. The exception is variable 1 which loads moderately on both factors. This is explained by the fact that this is a measure of Spearman's general factor to which both verbal and problem-solving abilities contribute. In fact this study is somewhat unusual in including a composite variable of this kind, but for pedagogic purposes it has the advantage of giving greater insight into the factor structure. There is a further rotation which brings this point out more clearly. If we were to rotate the axes orthogonally so that one axis goes through the point for variable 1 we should then have one general factor with all loadings positive and one bi-polar factor in which problem-solving variables have positive loadings and verbal

Table 3.5 Results of Oblimin rotation for Smith and Stanley's data

	Maximum likelihood		Principal axis (least squares)	
i	λ_{i1}	λ_{i2}	λ_{i1}	λ_{i2}
1	.39	.46	.50	−.37
2	−.01	.64	.62	.01
3	−.02	.89	.95	.08
4	−.02	.49	.44	−.01
5	1.01	−.06	−.05	−.99
6	.80	.03	.03	−.81

variables negative loadings. This second factor would then distinguish those whose strength is in language from those strong on reasoning.

Confirmatory Factor Analysis

We have presented factor analysis as an exploratory technique which sets out to express a many-dimensional data set in terms of a much smaller number of dimensions or factors. But we noted above that there may be prior evidence that $\mathbf{\Lambda}$ has a particular structure; for example, that the manifest variables can be divided into sub-sets each one of which depends primarily on a single factor. We shall then be interested only in solutions which conform to this pattern. The analysis and, in particular, the estimation should then take account of this special knowledge. Such an analysis is called *confirmatory factor analysis* and in some fields of application, psychology for example, it is the norm. The chief difference arises over the matter of rotation. Any particular pattern of zeros in a $\mathbf{\Lambda}$-matrix will be destroyed if the matrix is subject to rotation. The infinitely many possible solutions offered by exploratory analysis are no longer available. Instead we shall be looking for a solution with a pre-specified pattern for $\mathbf{\Lambda}$ which may (and usually will) entail correlated factors.

If we allow the y's in the linear model to be dependent with correlation matrix $\mathbf{\Phi}$ then instead of (3.4) we shall have

$$\mathbf{\Sigma} = \mathbf{\Lambda}\mathbf{\Phi}\mathbf{\Lambda}' + \boldsymbol{\psi}. \tag{3.45}$$

Clearly, the parameters are only identified if there are sufficient restrictions imposed upon them which will usually be in the form of elements of $\mathbf{\Lambda}$ which are zero.

There are two approaches to solving this more general problem. One is to proceed as before to find the maximum likelihood estimates but differentiating with respect to the free parameters only. The likelihood is as in (3.6) but with $\mathbf{\Sigma}$ given by (3.45). The partial derivatives with respect to $\mathbf{\Lambda}$ and $\boldsymbol{\psi}$ involve replacing (3.9) by

$$2\mathbf{\Omega}\mathbf{\Lambda}\mathbf{\Phi}, \tag{3.46}$$

(3.10) is unchanged and

$$-\frac{2}{n}\frac{\partial L}{\partial \mathbf{\Phi}} = 2\mathbf{\Lambda}'\mathbf{\Omega}\mathbf{\Lambda}. \tag{3.47}$$

The elements in these various matrices of partial derivatives corresponding to fixed parameters are set equal to zero. The zig-zag type of solution is no longer available but numerical optimization routines are available in the LISREL and COSAN packages.

The second approach is to fit an unrestricted model and then to estimate a rotation matrix **M** such that the pattern of the rotated matrix is as close to that required as possible. The correlation matrix for the factors will then be $\mathbf{\Phi} = \mathbf{MM}'$.

The determination of a suitable transformation is sometimes referred to as the Procrustes problem after the figure in Greek mythology who kept a bed into which his victims had to be fitted by various painful means. One may allow arbitrary rotations or restrict them to be orthogonal. The problem is a long-standing one to which contributions have been made by, among others, Green (1952), Cliff (1966), Schönemann (1966), Lawley and Maxwell (1971, Sections 6.5 and 6.6) and Brokken (1983). This approach does not appear to be widely used, though in many ways it seems preferable to fixing some parameters at zero. We would hardly expect individual loadings to be exactly zero and the method allows us to see how close we can get to the desired pattern without loss of fit. The Procrustes approach gives a better fit to the data at the price of a poorer reproduction of the desired pattern of loadings.

Confirmatory factor analysis raises other important questions of a non-statistical kind. For example, if it is really known *a priori* which sub-groups of manifest variables are related to which factors it may be better to fit models with fewer factors to each sub-group in the first place.

3.5 Factor scores

Having determined the dimensionality of the factor space and attempted an interpretation it may be necessary to go on and construct scales of measurement for the factors. If, for example, we have identified a factor designated "arithmetical ability" we may wish to construct an index which purports to measure that ability. In abstract terms our problem is to use the information contained in an individual's **x**-value to locate him in the factor space **y**. There has been a long and controversial debate about how best to do this. The main point has often been obscured by talk of "estimating" the factor scores as if they were parameters in the ordinary statistical sense. If we follow the approach of Chapter 1 in which the factors are random variables it is clear that what we have is a prediction problem and not one of estimation. The factors will be random variables *after* **x** has been observed, as well as before. Any attempt to locate an individual in the factor space must therefore be based on the posterior distribution of **y** given **x**. This point was argued in Bartholomew (1981) but had been anticipated by Dolby (1976).

To do this, of course, the prior distribution must be specified, and as this is essentially arbitrary, progress appears to be blocked. However, if we agree to suppose that a factor is scaled in such a way as to make its prior distribution normal, say, then we can proceed to predict the location of individuals on that scale. (In the next chapter we shall see that the ranking of predicted positions on that scale does not depend on the choice of prior distribution.)

In the case of the normal model we derived the posterior distribution of y in (1.13) and found it to be normal. We could thus predict an individual's $\mathbf{y}$ using the posterior mean, and our "factor scores" would then be

$$\begin{aligned} E(\mathbf{y} \mid \mathbf{x}) &= \mathbf{\Lambda}'\mathbf{\Sigma}^{-1}(\mathbf{x}-\boldsymbol{\mu}) \\ &= (\mathbf{I}+\mathbf{\Gamma})^{-1}\mathbf{\Lambda}'\boldsymbol{\psi}^{-1}(\mathbf{x}-\boldsymbol{\mu}). \end{aligned} \tag{3.48}$$

The equivalence of the two right-hand expressions follows by noting that

$$\begin{aligned} \mathbf{\Lambda}'\boldsymbol{\psi}^{-1}\mathbf{\Sigma} &= \mathbf{\Lambda}'\boldsymbol{\psi}^{-1}(\mathbf{\Lambda}\mathbf{\Lambda}'+\boldsymbol{\psi}) \\ &= (\mathbf{I}+\mathbf{\Gamma})\mathbf{\Lambda}'. \end{aligned}$$

The scores are thus linear functions of the x's the coefficients of which would have to be estimated. A measure of the variation of the y's about their expectations can be obtained from the diagonal elements of the dispersion matrix $(\mathbf{I}+\mathbf{\Gamma})^{-1}$. For the same reasons as set out below (3.40) the posterior variances will diminish as p increases. In other words, the more manifest variables we have the more precisely will the factors be determined. (For a more formal treatment of this matter see Kano (1984).)

Distribution-free Methods

If we are unwilling to make assumptions about the forms of the distributions in the linear model we can still aim to find functions of the x's which are, in some sense, as near as possible to the y's. In the case of the linear model it is sufficient to consider only linear functions as we now show. Let us consider the q linear functions $\mathbf{X}$ obtained by pre-multiplying $(\mathbf{x}-\boldsymbol{\mu})$ by a $q \times p$ matrix of the form $(\mathbf{B\Lambda})^{-1}\mathbf{B}$. Then from (3.1)

$$\mathbf{X} = (\mathbf{B\Lambda})^{-1}\mathbf{B}(\mathbf{x}-\boldsymbol{\mu}) = \mathbf{y} + (\mathbf{B\Lambda})^{-1}\mathbf{Be} \tag{3.49}$$

and hence we may write

$$X_j = y_j + u_j \quad (j = 1, 2, \ldots, q). \tag{3.50}$$

We could then regard X_j as a factor score for y_j since u_j is a random error independent of y_j. Since the transformation in (3.49) involves the arbitrary matrix **B** we shall seek to choose the transformation so that $\text{var}(u_j)$ is as small as possible for each j. In this way X_j is made as close as possible to y_j in a mean square sense.

Let $\mathbf{C} = (\mathbf{B\Lambda})^{-1}\mathbf{B}$; then

$$u_j = \sum_{h=1}^{p} c_{jh} e_h \quad \text{and} \quad \text{var}(u_j) = \sum_{h=1}^{p} c_{jh}^2 \psi_h. \tag{3.51}$$

The variance has to be minimized, for each j, subject to the constraint

$$\mathbf{C\Lambda} = \mathbf{I}. \tag{3.52}$$

Using Lagrange's method we minimize

$$\phi_j = \sum_{h=1}^{p} c_{jh}^2 \psi_h + \sum_{k=1}^{q} \mu_{jk} \sum_{h=1}^{p} c_{jk} \lambda_{hk} \tag{3.53}$$

where the μ's are undetermined multipliers. The partial derivatives are

$$\frac{\partial \phi_j}{\partial c_{jh}} = 2c_{jh}\psi_h + \sum_{k=1}^{q} \mu_{jk}\lambda_{hk} \quad (h = 1, 2, \ldots, p).$$

The derivatives vanish when

$$c_{jh} = -\tfrac{1}{2} \sum_{k=1}^{q} \mu_{jk}\lambda_{hk}/\psi_h$$

or, in matrix notation, when

$$\mathbf{C} = -\tfrac{1}{2}\mathbf{M\Lambda}'\boldsymbol{\psi}^{-1} \tag{3.54}$$

where $\mathbf{M} = \{\mu_{jk}\}$. Substituting into (3.52) we also have

$$-\tfrac{1}{2}\mathbf{M\Lambda}'\boldsymbol{\psi}^{-1}\mathbf{\Lambda} = \mathbf{I}. \tag{3.55}$$

Eliminating **M** between (3.54) and (3.55) we find

$$\mathbf{C} = \mathbf{\Gamma}^{-1}\mathbf{\Lambda}'\boldsymbol{\psi}^{-1}$$

whence

$$\mathbf{X} = \mathbf{\Gamma}^{-1}\mathbf{\Lambda}'\boldsymbol{\psi}^{-1}(\mathbf{x} - \boldsymbol{\mu}) \quad \text{and} \quad D(\mathbf{u}) = \mathbf{\Gamma}^{-1}. \tag{3.56}$$

These are known as Bartlett's scores (Bartlett 1937, 1938) and the advantage often claimed for them is that $E(\mathbf{x} \mid \mathbf{y}) = \mathbf{y}$ as (3.50) shows. However, an expectation of **x** conditional on **y** is hardly relevant when it is **x** that is known and **y** that is to be predicted. Of more interest is the comparison with (3.48). The only difference is that $(\mathbf{I} + \mathbf{\Gamma})^{-1}$ in (3.48) is replaced by $\mathbf{\Gamma}^{-1}$ in (3.56). If $\mathbf{\Gamma}$ is diagonal the effect of this difference is merely to introduce different scalings of the X's which is of no practical significance. The distribution-free argument can thus be viewed as establishing the result of (3.48) on a broader basis. However, a slightly different approach leads to scores which are identical with those of (3.48) as we now show.

In this method, due to Thomson (1939), we choose X_j to be as close to y_j as possible in the sense of minimizing

$$\phi_j = E(X_j - y_j)^2 \tag{3.57}$$

where the expectation is with respect to both $\mathbf{y}$ and $\mathbf{x}$. Here we have

$$\begin{aligned}\frac{\partial \phi_j}{\partial c_{jh}} &= 2E(x_h - \mu_h)\left(\sum_{k=1}^{p} c_{jk}(x_k - \mu_k) - y_j\right) \\ &= 2\left(\sum_{k=1}^{p} c_{jk}\sigma_{kh} - \lambda_{hj}\right).\end{aligned}$$

Setting the derivatives equal to zero and collecting together the results for all j and h gives

$$\mathbf{C\Sigma} = \mathbf{\Lambda}'$$

whence

$$\mathbf{C} = \mathbf{\Lambda}'\Sigma^{-1} = (\mathbf{I} + \mathbf{\Gamma})^{-1}\mathbf{\Lambda}'\boldsymbol{\psi}^{-1}. \tag{3.58}$$

When derived in this way the scores are known as "regression" scores. In this case it is *not* true that $E(\mathbf{x} \mid \mathbf{y}) = \mathbf{y}$, but for the reason given above this is not a relevant objection.

3.6 Heywood cases

All the methods of estimating the parameters of the factor model involve minimizing a measure of the distance between $\mathbf{S}$ and $\mathbf{\Sigma}$. However, the parameter space is restricted by the condition $\boldsymbol{\psi} \geqslant \mathbf{0}$ so the usual procedure of setting all the partial derivatives equal to zero may yield a solution with a negative ψ_i. In such cases the minimum we seek will lie on a boundary of the admissible region at a point where one or more of the ψ's is zero. When this happens we have what is called a Heywood case after Heywood (1931). In practice this will be recognized either by the appearance of a negative estimate or by the convergence of an estimate to zero, depending on the algorithm used.

There is no inconsistency in the occurrence of a zero residual variance and, taken at its face value, it would simply mean that the variation of the manifest variable in question was wholly explained by the latent variables. In practice this rarely seems plausible and the rather frequent occurrence of Heywood cases has caused a good deal of unease among practitioners. Taken with the fact that it is known that zero estimates can easily arise when the true parameter values are not zero, there is good reason for not taking such estimates at their face value.

There is a good deal of evidence, mainly of an empirical kind, about the circumstances under which Heywood cases are likely to occur. Much of this is based on simulation studies where there is no question of the model itself being invalid. The results are widely scattered in the literature but we have drawn heavily in what follows on van Driel (1978), Anderson and Gerbing (1984), Boomsma (1985) and Fachel (1986).

If a Heywood case arises when the data conform to the linear factor model it will almost certainly be the result of sampling error. The key factor is therefore sample size. For a model with positive ψ's the probability of a Heywood case tends to zero as n tends to infinity. We are therefore dealing with a small sample phenomenon. The risk of a Heywood case depends on other factors mentioned below, but as a rough guide it appears that it is high with sample sizes of 100 or less and low with samples of 500 or more.

For a given sample size the risk decreases as p, the number of variables, increases. Obviously the risk will also be greater if one or more ψ's is very small, but this is not something which would be known in advance. Nevertheless, one can obtain some clues from the data. For example, in (3.40) and the remarks that followed we saw that an estimated upper bound for ψ_i could be obtained and, in particular, that ψ_i would be small if x_i was highly correlated with any other variable. The presence of one or more high correlations is therefore indicative of a potential Heywood case.

Perhaps the commonest cause of Heywood cases is the attempt to extract more factors than are present. This is readily demonstrated by simulation but might have been anticipated on the grounds that artificially inflating the communality forces the residuals towards zero.

Once the causes of Heywood cases are understood, the ways of dealing with them become clearer. At the stage of designing an enquiry one should aim for a large sample with a good number of variables. But in selecting variables it is important to avoid introducing new variables which add little to those already there. This will merely create high correlations without contributing significantly to the information about the latent variables.

At the analysis stage the options are more limited. Over-factoring can be avoided by paying careful attention to the various criteria suggested in Section 3.2. If a highly correlated pair of variables appears to be the cause, one of them can be dropped with little loss. However, it does not follow that dropping a variable which is implicated in a Heywood case is always advisable. If it arises because of a single small residual variance we should then be omitting one of the more valuable variables which was a relatively pure indicator of the latent variables. In any case, experience shows that such a course often leads to a Heywood case involving another variable which defeats the object.

If the foregoing precautions fail there are at least three courses still open:

(a) We can use a Bayesian approach like that suggested by Martin and McDonald (1975). Here we maximize not the likelihood but the posterior density, using a prior distribution for the ψ's which assigns zero probability to negative values. Martin and McDonald propose a form which is both tractable and plausible in that it implies a distribution which is almost uniform except that it decreases to zero at the point $\psi_i = 0$.
(b) We may stop the iteration at some arbitrarily small value of ψ_i such as 0.05 or 0.01. In effect, this is a special case of (a) with a uniform prior on the interval $(\varepsilon, 1)$ where ε is the chosen cut-off point. It may be justified on the grounds that the likelihood (or other distance measure) will be very close to its optimum and all that matters for purposes of interpretation is to know that ψ_i is "small".
(c) A third method which shows promise but needs further investigation rests on the following simple idea. If a variable x_i has a small ψ_i we could increase the latter by adding to x_i an independent random variable with known variance σ^2. Denoting the new variable by x_i' its variance would then be

$$\text{var}(x_i') = \sum_{j=1}^{q} \lambda_{ij}^2 + \psi_i + \sigma^2.$$

If x_i' were used instead of x_i in estimating the parameters, we would obtain an estimate of $\psi_i + \sigma^2$. Knowing σ^2 we could then obtain an estimate of ψ_i by subtraction. It is not, in fact, necessary to add the artificial variable to each value of x_i; we could simply add σ^2 to the appropriate diagonal element of $\mathbf{S}$.

If the analysis is carried out on the correlation matrix the effect of replacing x_i by x_i' is to multiply the off-diagonal elements in the ith row of the sample correlation matrix by $(1+\sigma^2)^{-\frac{1}{2}}$. The relationship between the parameters of the original model and the modified one is then given by

$$\lambda_{ij}^* = \lambda_{ij}'(1+\sigma^2)^{\frac{1}{2}}, \quad \psi_i^* = 1 - (1+\sigma^2) \sum_{j=1}^{q} \lambda_{ij}'^2 \tag{3.59}$$

where the prime denotes the modified loadings. The result holds for any value of σ^2 but it seems reasonable to choose a value just large enough to avoid the occurrence of a Heywood case. In practice $\sigma^2 = 1$ seems to be the right order of magnitude. It should be noted that the results on the standard errors of the estimates and on goodness of fit will be invalidated by this device.

CHAPTER 4

Foundations

4.1 Introduction

The latent variable models in common use have usually been introduced in a natural but somewhat arbitrary way. For example, the standard linear model of factor analysis discussed in Chapter 2 follows the pattern of the general linear model of statistics in supposing that the factors exert their effect on the distribution of the x's only through the mean. We have seen in Chapter 1 that such a model can be justified in terms of normal variables, but it is also clear from the same treatment that the class of possible models is very large. Faced with such a variety we need some guiding principles of choice. Ideally we would like to narrow down the class of possible models to include just sufficient for practical purposes. It is the purpose of this chapter to set out a general approach to the construction of latent variable models based on the statistical idea of sufficiency following Bartholomew (1984a).

The new approach will first enable us to see the models of the last two chapters in a broader perspective and to understand better the nature of the arbitrariness which they involve. It will also suggest alternative ways of interpreting the results of factor analysis and related methods. Secondly, the approach will provide a basis for developing new models especially for non-normal variables. Categorical variables provide an important example of non-normality which can be accommodated within the general framework. The remaining chapters are concerned with working out the implications of this remark for binary and polytomous data.

A unifying feature of the treatment is the central place given to what will be called components. These are not principal components but the use of the term is suggested by the fact that the role which they play in factor analysis is similar to that of the traditional principal components. They are, in fact, the set of minimal sufficient statistics arising from our general formulation. Often they will be linear in $\mathbf{x}$, in which case the interpretation can be based on the magnitude and signs of the coefficients. Although the new methods are equivalent to the old in the sense that one can be derived from the other, the emphasis is shifted away from the latent variables, which cannot be directly observed, to the components which can. In so far as these can be given a substantive interpretation in their own right, the model itself is less critical since the results of the analysis can be given meaning without direct reference to the model from which it has been derived.

We begin, as in Chapter 1, by treating the latent variables as continuous, but in Section 4.5 we shall extend the theory to take in categorical latent variables.

4.2 The sufficiency principle

We saw in Chapter 1 that the specification of a model requires the choice of a set of probability distributions. One of these is the prior distribution of the latent variables, $h(\mathbf{y})$, which is essentially arbitrary. Later we shall see that the choice is not critical but, in any event, the main result of this section is independent of the prior distribution. Secondly there is the set of conditional distributions $\{g_i(x_i \mid \mathbf{y})\}$. These are not directly observable, since we cannot fix $\mathbf{y}$, and so it might seem that there could be little rational basis for the choice. Fortunately it turns out that the objectives of our analysis place quite severe restrictions on the class of conditional distributions which it is reasonable to consider.

We recall first that our prime objective is to explain the interdependences of the p x's in terms of a much smaller number, q, of latent variables, $\mathbf{y}$. The x's themselves obviously contain all the information about $\mathbf{y}$, but it might be possible to convey this more economically in terms of a smaller number of summary statistics. We could hardly expect this set to have fewer than q members, but it would be particularly appropriate if it contained exactly q members. We would then have information about the q-dimensional space of $\mathbf{y}$ conveyed by a vector of observable quantities of exactly the same dimension. This leads to the question: For what class of conditional distributions of x_i given $\mathbf{y}$ is such a reduction possible? The answer is that it is the class for which the posterior distribution $h(\mathbf{y} \mid \mathbf{x})$ depends on $\mathbf{x}$ only through q functions of $\mathbf{x}$. This will be so if

$$\prod_{i=1}^{p} g_i(x_i \mid \mathbf{y}) \tag{4.1}$$

is a function of these same q functions. In the language of statistical estimation theory, we are asking that the joint density of (4.1) should admit a q-dimensional minimal sufficient statistic for $\mathbf{y}$ (which, being fixed, may be regarded as a parameter). The class of $g_i(x_i \mid \mathbf{y})$'s for which this is the case has been determined by Barankin and Maitra (1963, Theorems 5.1, 5.2 and 5.3). They give the necessary and sufficient conditions which amount to the requirement that at least $(p-q)$ of the g_i's shall be of exponential type, that is that

$$g_i(x_i \mid \mathbf{y}) = F_i(x_i)G_i(\mathbf{y}) \exp \sum_{j=1}^{q} u_{ij}(x_i)\phi_j(\mathbf{y}). \tag{4.2}$$

In most cases, in practice, we shall want to assume that all our x's have a distribution of the same form, and (4.2) tells us the class from which our selection must be made. However, there is nothing in the general theory to prevent us from having different members of the exponential family for different x's. (The significance of the requirement that only $(p-q)$ of the distributions need be of this form can be appreciated by considering the extreme case $p=q$; $\mathbf{x}$ is then itself a minimal statistic whatever the form of the distribution.) Since, in practice, q is very much smaller than p there is little loss of generality in requiring that all y's shall be of the form (4.2) and this we shall do. In this case

$$h(\mathbf{y} \mid \mathbf{x}) = \frac{h(\mathbf{y}) \prod_{i=1}^{p} F_i(x_i)G_i(\mathbf{y}) \exp \sum_{j=1}^{q} X_j\phi_j(\mathbf{y})}{\int h(\mathbf{y}) \prod_{i=1}^{p} F_i(x_i)G_i(\mathbf{y}) \exp \sum_{j=1}^{q} X_j\phi(\mathbf{y})\, \mathrm{d}\mathbf{y}} \tag{4.3}$$

where $X_j = \sum_{i=1}^{p} u_{ij}(x_i)$. The product $\prod_{i=1}^{p} F_i(x_i)$ cancels and we are left with an expression which depends on $\mathbf{x}$ only through the q components $\{X_j\}$.

The exponential family is broad enough to cover most practical needs and hence we shall be able to achieve our aim of finding a q-dimensional reduction of the data. Notice that we are not saying that g_i must belong to the exponential family but only that if it does not then it will not be possible to achieve the reduction in dimensionality without some loss of information.

We emphasize again that nothing in the argument so far places any restriction on the distribution of the y's. We can use this fact to simplify (4.3) by the transformation

$$z_j = \phi_j(\mathbf{y}) \quad (j = 1, 2, \ldots, q). \tag{4.4}$$

The choice of z's rather than y_j's is of no theoretical significance but it has the convenient property of leading to linear models.

There are two ways in which we could use these results to derive latent variable models. The first is of little relevance for our main purpose but it serves to throw into sharper relief the essential character of the second. It may be illustrated by the following example for $q = 2$. Let

$$g_i(x_i \mid \mathbf{y}) = \frac{1}{\sqrt{2\pi}} \frac{1}{\sqrt{y_2}} \exp - \frac{1}{2} \frac{(x_i - y_1)^2}{y_2}; \tag{4.5}$$

y_1 is thus the mean and y_2 the variance of x_i for all i. This may be written in the form (4.2) with

$$F_i(x_i) = 1, \quad G_i(\mathbf{y}) = \frac{1}{\sqrt{y_2}} \exp\left\{ - \frac{1}{2} \frac{y_1^2}{y_2} \right\}$$

$$u_{i1}(x_i) = x_i, \quad u_{i2}(x_i) = x_i^2, \quad \phi_1(\mathbf{y}) = y_1/y_2, \quad \phi_2(\mathbf{y}) = -1/2y_2$$

with the new latent variables of (4.4) given by

$$z_1 = y_1/y_2 \quad \text{and} \quad z_2 = -1/2y_2.$$

The two components are thus $X_1 = \sum_{i=1}^{p} x_i$ and $X_2 = \sum_{i=1}^{p} x_i^2$ which contain all the information in the x's about the two latent variables z_1 and z_2. We could work out the posterior density from (4.3) and express its mean and covariance matrix as functions of X_1 and X_2. The model could also be written

$$x_i = y_1 + e_i, \quad (i = 1, 2, \ldots, p) \tag{4.6}$$

where $e_i \frown N(0, y_2)$. One latent variable relates to the mean of x_i and the other to the variance of the error term.

Although this is a simple model which requires no fitting it is not the kind of model which factor analysts commonly require. The second way of proceeding allows all the latent variables to influence a single parameter. Returning to (4.2), consider the sub-class of distributions for which

$$u_{ij}(x_i) = \alpha_{ij} u_i(x_i) \quad \text{for all } i \text{ and } j \tag{4.7}$$

then

$$g_i(x_i \mid \mathbf{z}) = F_i(x_i) G_i(\theta_i) \exp \theta_i u_i(x_i) \tag{4.8}$$

where $\theta_i = \sum_{j=1}^{q} \alpha_{ij} z_j$. This is a one-parameter distribution of the exponential family, and θ_i is known as the *natural* or *canonical* parameter. *Our model is then that θ_i is a linear function of the q latent variables $z_1, z_2, \ldots, z_q$.* To see how this works out in the normal case, suppose $x_i \frown N(\mu_i, \sigma_i^2)$. There are two ways in which this distribution can be expressed in the form (4.8):

$$\text{(i)} \ \ \theta_i = \mu_i/\sigma_i, \quad u_i(x_i) = x_i,$$

$$\text{(ii)} \ \ \theta_i = -1/2\sigma_i^2, \quad u_i(x_i) = x_i^2.$$

In the first case σ_i would be a nuisance parameter and in the second μ_i. The two models could then be written

$$\text{(i)}\ x_i = \sigma_i \sum_{j=1}^{q} \alpha_{ij} z_j + e_i, \quad \text{where } E(e_i) = 0 \text{ and } \operatorname{var}(e_i) = \sigma_i^2 \tag{4.9}$$

$$\text{(ii)}\ x_i = \mu_i + e_i \quad \text{where } E(e_i) = 0,\ \operatorname{var}(e_i) = -1 \Big/ \left(2 \sum_{j=1}^{q} \alpha_{ij} z_j\right). \tag{4.10}$$

The first of these will be recognized as equivalent to the standard linear factor model with which we shall be primarily concerned, but we have considered this example in some detail to emphasize the great variety of models implicit in our formulation. Model (ii), for example, would allow one to investigate the effect of latent variables on the dispersion of the manifest variables. This model also alerts us to the fact that the choice of the linear model may entail restrictions on the range and sign both of the z's and the α's since a variance must be positive. In the case of (i) we usually require the latent variables to have zero means, in which case we must introduce a constant onto the right-hand side.

For the general model based on (4.8) with

$$\theta_i = \sum_{j=1}^{q} \alpha_{ij} z_j \tag{4.11}$$

the components are given by

$$X_j = \sum_{i=1}^{p} \alpha_{ij} u_i(x_i) \quad (j = 1, 2, \ldots, q) \tag{4.12}$$

or, in matrix notation,

$$\mathbf{X} = \mathbf{A}'\mathbf{u} \tag{4.13}$$

where $\mathbf{A} = \{\alpha_{ij}\}$ and $\mathbf{u}' = \{u_1(x_1), u_2(x_2), \ldots, u_p(x_p)\}$. In the normal model (i) above the components are

$$X_j = \sum_{i=1}^{p} (\alpha_{ij}/\sigma_i) x_i \quad (j = 1, 2, \ldots, q). \tag{4.14}$$

Example

As a further illustration we take a very simple but unfamiliar case. If the x's were durations of some kind it might be reasonable to suppose that they had an exponential distribution with

$$g_i(x_i \mid \theta_i) = \theta_i \exp(-\theta_i x_i) \quad (i = 1, 2, \ldots, p). \tag{4.15}$$

This is a member of the one-parameter exponential family with

$$F_i(x_i) = 1, \quad G_i(\theta_i) = \theta_i, \quad \text{and} \quad u_i(x_i) = -x_i.$$

The model would then be

$$\theta_i = \sum_{j=1}^{q} \alpha_{ij} z_j \quad (i = 1, 2, \ldots, p)$$

where, again, restrictions have to be imposed to ensure that $\theta_i \geqslant 0$. The theory tells us that the rate parameter (not the mean) should depend linearly on the latent variables.

4.3 Interpretation

We can approach the matter of interpretation at two levels. One is to examine the components and to ask what it is they are measuring. This can be done without any explicit reference to the model which has given rise to them. The second approach is to establish a link between the components and the underlying factors and then to express the interpretation in terms of the factors. As we shall see, there are some general results applicable to all models but individual models have their own special features. In the following chapters we shall find that this is true of problems with categorical manifest variables. Here we shall illustrate the ideas on the normal linear model. This exhibits most of the features which have to be considered and also enables us to compare the new approach with the traditional methods already set out in Chapter 3.

Interpretation is facilitated by the fact that the components are always linear combinations. Thus since

$$X_j = \sum_{i=1}^{p} \alpha_{ij} u_i(x_i) \tag{4.16}$$

it is the relative magnitudes and the signs of the α's which are the key to interpretation. A variable $u_i(x_i)$ with a large α is relatively important in determining the value of X_j and one would then ask what the variables with large coefficients of the same sign have in common. It should be noted that $u_i(x_i)$ may involve nuisance parameters as it does in the normal case for which

$$X_j = \sum_{i=1}^{p} \alpha_{ij} x_i / \sigma_i.$$

If we were to attempt to interpret X_j by reference to the coefficients $\{\alpha_{ij}/\sigma_i\}$ of x_i the result would be affected by the scales of the x_i's since σ_i depends on the scale of x_i but α_{ij} does not. This objection does not apply to the use of the α's alone, but then the scaled variable x_i/σ_i lacks an immediate interpretation.

A better alternative is to write X_j in the form

$$X_j = \sum_{i=1}^{p} (\alpha_{ij}\sigma_i(x_i)/\sigma_i)(x_i/\sigma(x_i)) \tag{4.17}$$

and interpret the component by reference to the coefficients

$$\frac{\alpha_{ij}\sigma_i(x_i)}{\sigma_i}, \quad (i = 1, 2, \ldots, p) \tag{4.18}$$

which are the weights applied to the standardized x's. There is a close analogy here with the interpretation of principal components where the variables are usually standardized. We now write these coefficients in a more familiar form. From (4.9) $\lambda_{ij} = \sigma_i\alpha_{ij}$ and $\sigma_i = \psi_i$, $\sigma(x_i) = \left(\sum_{j=1}^{q} \lambda_{ij}^2 + \psi_i\right)^{\frac{1}{2}}$, so (4.18) becomes

$$\frac{\lambda_{ij}}{\psi_i}\left(\sum_j \lambda_{ij}^2 + \psi_i\right)^{\frac{1}{2}}. \tag{4.19}$$

If we now express this in terms of the scale-invariant parameters

$$\lambda_{ij}^* = \lambda_{ij}\Big/\left(\sum_j \lambda_{ij}^2 + \psi_i\right)^{\frac{1}{2}} \quad \text{and} \quad \psi_i^* = \psi_i\Big/\left(\sum_j \lambda_{ij}^2 + \psi_i\right)$$

(4.18) finally becomes

$$\lambda_{ij}^*/\psi_i^*. \tag{4.20}$$

If we are to interpret the components consistently with the way we interpret principal components we shall therefore use the "loadings" or "weights" as given by (4.20). In the same notation

$$\alpha_{ij} = \lambda_{ij}^*/\sqrt{\psi_i^*}. \tag{4.21}$$

We thus have three possible sets of coefficients to use in the interpretation of the components:

(i) the λ_{ij}^*'s of the traditional treatment given in Chapter 3;
(ii) the α_{ij}'s of (4.21) suggested by the general form of the components as given in (4.16);
(iii) the coefficients of (4.20).

It is clear that the signs of the various coefficients for any j will be the same and that any difference in their relative values will depend on the degree of variation in the ψ's. The position will be illustrated using an example below, but first we note that (i) above can be interpreted in a different way to that of Chapter 3. We saw there that, in the case of orthogonal factors, λ_{ij}^* was the correlation between x_i and y_j. Here, we show that it is proportional to the correlation between x_i and X_j. Hence we can use it in exactly the same way to interpret the components as in the case of factors.

We first need to establish a preliminary result which is of independent interest, namely that the components are uncorrelated if and only if the matrix $\mathbf{\Gamma} = \mathbf{\Lambda}'\mathbf{\psi}^{-1}\mathbf{\Lambda} = \mathbf{A}'\mathbf{A}$ is diagonal. Writing

$$\mathbf{X} = \mathbf{\Lambda}'\mathbf{\psi}^{-1}\mathbf{x}$$

$$D(\mathbf{X}) = \mathbf{\Lambda}'\psi^{-1}D(\mathbf{x})\mathbf{\psi}^{-1}\mathbf{\Lambda} = \mathbf{\Lambda}'\mathbf{\psi}^{-1}(\mathbf{\Lambda}\mathbf{\Lambda}' + \mathbf{\psi})\mathbf{\psi}^{-1}\mathbf{\Lambda}$$
$$= \mathbf{\Gamma}^2 + \mathbf{\Gamma}$$

and the result is proved. This is precisely the condition satisfied by the maximum likelihood estimators obtained by the method of Chapter 3. Since uncorrelated components will usually be easier to interpret we may regard this result as a justification for what, in the usual approach, is regarded as an arbitrary constraint.

We now find the matrix of correlations between the variables and components.

$$E(\mathbf{x} - \mathbf{\mu})(\mathbf{X} - E(\mathbf{X}))' = D(\mathbf{x})\mathbf{\psi}^{-1}\mathbf{\Lambda} = (\mathbf{\Lambda}\mathbf{\Lambda}' + \mathbf{\psi})\mathbf{\psi}^{-1}\mathbf{\Lambda} = \mathbf{\Lambda}(\mathbf{\Gamma} + \mathbf{I})$$

and

$$\operatorname{corr}(\mathbf{x}, \mathbf{X}') = \{\operatorname{diag}(\mathbf{\Lambda}\mathbf{\Lambda}' + \mathbf{\psi})\}^{-\frac{1}{2}}\mathbf{\Lambda}(\mathbf{\Gamma} + \mathbf{I})\{\operatorname{diag}(\mathbf{\Gamma}^2 + \mathbf{\Gamma})\}^{-\frac{1}{2}}.$$

If $\mathbf{\Gamma}$ is diagonal this becomes

$$\operatorname{corr}(\mathbf{x}, \mathbf{X}') = \mathbf{\Lambda}^*(\mathbf{I} + \mathbf{\Gamma}^{-1})$$

or

$$\operatorname{corr}(x_i, X_j) = \lambda_{ij}^*(1 + \Gamma_j^{-1}) \tag{4.22}$$

where Γ_j is the jth diagonal element of $\mathbf{\Gamma}$. Thus for given j the correlations are proportional (and usually close) to the λ_{ij}'s. We repeat that this interpretation will only be available if the components are uncorrelated.

Example

Suppose the correlation matrix is

1	.8636	.6792	.5099	.6347	.6396
	1	.6738	.5164	.5738	.6232
		1	.3274	.4165	.4614
			1	.6751	.7377
				1	.8384
					1

Table 4.1

i	1	2	3	4	5	6
λ^*_{i1}	.882	.863	.658	.704	.822	.859
λ^*_{i2}	−.302	−.336	−.324	.329	.357	.365
ψ^*_i	.133	.143	.460	.394	.191	.133

If the standard normal factor model is fitted by maximum likelihood assuming two factors (for reasons which will appear later) the parameter estimates are obtained as in Table 4.1. The pattern of the loadings is fairly typical suggesting a first "general" factor and a second "bi-polar" factor. All manifest variables contribute substantially to the first factor, and each member of each of the sub-groups (1, 2, 3) and (4, 5, 6) contributes nearly equally to the second factor.

If we adopt the perspective of the new approach we have the three options for interpretation described above. Starting with that using the correlations between the x's and the components using (4.22), we find

$$\Gamma_1 = 22.344, \quad \Gamma_2 = 3.646$$

and

$$(1+\Gamma_1^{-1})^{\frac{1}{2}} = 1.022, \quad (1+\Gamma_2^{-1})^{\frac{1}{2}} = 1.129.$$

Multiplying the first row of Table 4.1 by 1.022 and the second by 1.129 raises the correlations slightly, but the interpretation we give to X_1 and X_2 would be essentially the same as to y_1 and y_2 using the traditional method. The component coefficients for the other two methods are given in Table 4.2.

In this example the residual variances are far from equal and so the rows of Table 4.2 are not proportional to those of Table 4.1. However, the main lines of the interpretation arrived at above remain. Using either set of coefficients, the general and bi-polar components still emerge. The difference lies in the greater variation in the contributions which the individual x's make to the value of the component. We

Table 4.2 Alternative component coefficients for the example

i	1	2	3	4	5	6
λ^*_{i1}/ψ^*_i	6.643	6.015	1.431	1.786	4.305	6.469
λ^*_{i2}/ψ^*_i	−2.271	−2.341	−.704	.834	1.871	2.749
α_{i1}	2.418	2.282	.970	1.122	1.881	2.355
α_{i2}	−.828	−.889	−.478	.524	.817	1.001

$(\alpha_{ij} = \lambda^*_{ij}/(\psi^*_j)^{\frac{1}{2}})$

would judge x_1, x_2, and x_6 to be relatively more important than x_3 and x_4 in deciding what X_1 measures.

The argument used above showing that the coefficients in the upper part of Table 4.2 are the weights of the standardized x's provides strong grounds for preferring this option. Against this may be set the practical argument that ψ_i^* may be very small, or even zero, for some i, in which case the corresponding coefficient may be very large and poorly determined. However, the force of this is diminished by noting that in this circumstance the variable in question is virtually uncontaminated by error and little would be lost by using that variable alone.

Rotation

We saw in Chapter 3 that the interpretation of factors could sometimes be facilitated by making a linear transformation of the factor space. Similar arguments may be used for transforming the components. The minimal sufficient set $\mathbf{X}$ which we have been seeking to interpret is not unique. Any one-to-one transformation $\mathbf{X} \rightarrow \mathbf{X}^*$, say, will preserve all the information about $\mathbf{y}$, and if we found $\mathbf{X}^*$ easier to interpret we should be entitled to use it instead. Since $\mathbf{X}$ is itself a linear function it is natural to consider linear transformations and to consider the family of components given by

$$\mathbf{X}^* = \mathbf{M}\mathbf{X}. \tag{4.23}$$

Components are likely to be easier to interpret if they load heavily on relatively few x's, and so we might seek to choose $\mathbf{M}$ to achieve this end. No new theory is needed in the normal case because there is a close relationship between the transformations needed for components and those for factors. In Chapter 3 we saw that the model with loadings $\mathbf{\Lambda}$ and factors $\mathbf{y}$ was indistinguishable from one with loadings $\mathbf{\Lambda M}$ and factors $\mathbf{M}^{-1}\mathbf{y}$. The components do not depend on the distribution of the factors, so the only effect of the transformation on the components will be to replace $\mathbf{\Lambda}$ by $\mathbf{\Lambda M}$. The new components will therefore be

$$\mathbf{X}^* = \mathbf{M}'\mathbf{\Lambda}'\boldsymbol{\psi}^{-1}\mathbf{x} = \mathbf{M}'\mathbf{X}. \tag{4.24}$$

If $\mathbf{M}$ is orthogonal so that $\mathbf{M}^{-1} = \mathbf{M}'$ the same transformation is thus applied to the components as to the factors.

In practice the object of rotation is usually to produce, as nearly as possible, a desired pattern of elements in the transformed $\mathbf{\Lambda}$-matrix. It is clear from (4.24) that since $\boldsymbol{\psi}$ is diagonal the pattern of elements in $\mathbf{\Lambda}'\boldsymbol{\psi}$ will be the same as in $\mathbf{\Lambda}'$. Any matrix $\mathbf{M}$ which produces simple structure, say, in the old approach will produce a similar simple

Table 4.3 **Transformed component weights for the example**

i	1	2	3	4	5	6
First component	6.559	6.296	1.682	−0.047	0.062	0.172
Second component	0.650	0.228	−0.132	2.005	4.657	6.918

structure in the coefficients of the components. This is illustrated in Table 4.3 which results from applying the (non-orthogonal) transformation matrix

$$\mathbf{M} = \begin{bmatrix} .559 & .531 \\ -1.253 & 1.267 \end{bmatrix}$$

to the weights in the top half of Table 4.2. The result was obtained using the Oblimin rotation of SPSS with delta = 0 but could easily be found graphically (non-zero values of "delta" place a limit on the degree of obliqueness of the transformed axes). This transformation clearly associates the first component with the first three variables and the remaining variables with the second component. The correlation coefficient of the two components is 0.721.

It is usual to distinguish between orthogonal and non-orthogonal rotations. If $\mathbf{M}$ is orthogonal the transformed factors are still orthogonal, but this is not, in general, true of the components, so the advantage which orthogonality offers for interpretation does not carry over. On the other hand, there are transformations of the components which yield orthogonal components but which do not correspond to orthogonal transformations of the factor space. In fact any matrix $\mathbf{M}$ of the form $\mathbf{N}(\mathbf{\Gamma}^2 + \mathbf{\Gamma})^{-\frac{1}{2}}$ where $\mathbf{N}$ is orthogonal yields orthogonal components if applied to components for which $\mathbf{\Gamma}$ is diagonal. Thus if

$$\mathbf{X}^* = \mathbf{N}(\mathbf{\Gamma}^{2+}\mathbf{\Gamma})^{-\frac{1}{2}}\mathbf{X}$$

then

$$\begin{aligned} D(\mathbf{X}^*) &= \mathbf{N}(\mathbf{\Gamma}^2 + \mathbf{\Gamma})^{-\frac{1}{2}} D(\mathbf{X})(\mathbf{\Gamma}^2 + \mathbf{\Gamma})^{-\frac{1}{2}}\mathbf{N}' \\ &= \mathbf{N}\mathbf{N}' = \mathbf{I}. \end{aligned}$$

We can regard this transformation as first scaling the components to have equal variance and then applying an orthogonal rotation. One could devise methods of rotation similar to varimax to select an $\mathbf{N}$ which leads to an approximation to simple structure.

Relationship between Components and Factors

So far our discussion of interpretation has been almost entirely at the first of the levels mentioned earlier. Essentially we have been asking what a component can be supposed to be measuring rather than what relationship it bears to the underlying factors of the model. We now proceed to discuss interpretation at the second level beginning with the linear model.

There is a very simple relationship between the components and the canonical factors (those that correspond to making $\mathbf{\Gamma}$ diagonal). If we start with the usual linear model

$$\mathbf{x} = \boldsymbol{\mu} + \mathbf{\Lambda}\mathbf{y} + \mathbf{e}$$

and pre-multiply by $\mathbf{\Lambda}'\boldsymbol{\psi}^{-1}$ we obtain

$$\mathbf{X} = \mathbf{\Lambda}'\boldsymbol{\psi}^{-1}\mathbf{x} = \mathbf{\Lambda}'\boldsymbol{\psi}^{-1}\boldsymbol{\mu} + \mathbf{\Gamma}\mathbf{y} + \mathbf{\Lambda}'\boldsymbol{\psi}^{-1}\mathbf{e} \tag{4.25}$$

or

$$X_j = \text{constant} + \Gamma_j y_j + \text{“error”} \quad (j = 1, 2, \ldots, q).$$

This shows that the only factor on which X_j depends is y_j and this is the essential link between the component and the factor. Notice that it does not depend on the form of the distribution either of y_j or of $\mathbf{e}$. How good an indicator X_j is depends on the relative sizes of the factor and error parts of (4.25). This can be expressed in terms of variances since

$$\operatorname{var}(X_j) = \Gamma_j^2 + \{\mathbf{\Lambda}'\boldsymbol{\psi}^{-1}\boldsymbol{\psi}\boldsymbol{\psi}^{-1}\mathbf{\Lambda}\}_{jj} = \Gamma_j^2 + \Gamma_j.$$

The proportion of the variance of X_j attributable to the factor is thus $\Gamma_j^2/(\Gamma_j^2 + \Gamma_j) = \Gamma_j/(\Gamma_j + 1)$. For the numerical example used above, $\Gamma_1 = 22.344$, $\Gamma_2 = 3.646$, so 95.7% of the variation of X_1 is explained by the first factor and 78.5% of X_2 by the second factor.

Another way of looking at the relationships is through the correlations of the factors and components.

$$\begin{aligned}
\operatorname{corr}(\mathbf{y}, \mathbf{X}') &= E(\mathbf{y}\mathbf{x}'\boldsymbol{\psi}^{-1}\mathbf{\Lambda})\{\operatorname{diag}(\mathbf{\Gamma}^2 + \mathbf{\Gamma})\}^{-\frac{1}{2}} \\
&= E\{\mathbf{y}(\boldsymbol{\mu} + \mathbf{\Lambda}\mathbf{y} + \mathbf{e})'\boldsymbol{\psi}^{-1}\mathbf{\Lambda}\}\{\operatorname{diag}(\mathbf{\Gamma}^2 + \mathbf{\Gamma})\}^{-\frac{1}{2}} \\
&= \mathbf{\Gamma}\{\operatorname{diag}(\mathbf{\Gamma}^2 + \mathbf{\Gamma})\}^{-\frac{1}{2}};
\end{aligned} \tag{4.26}$$

or

$$\operatorname{corr}(y_j, X_j) = \{\Gamma_{jj}/(\Gamma_{jj} + 1)\}^{\frac{1}{2}}. \tag{4.27}$$

The relationship between components and factors which results from transformation can be obtained in a similar fashion. Thus if we transform to $\mathbf{z} = \mathbf{M}^{-1}\mathbf{y}$ with corresponding components $\mathbf{X}^* = \mathbf{M}'\mathbf{X}$ we find

$$\operatorname{corr}(\mathbf{z}, \mathbf{X}^{*\prime}) = \{\operatorname{diag} D(\mathbf{z})\}^{-\frac{1}{2}} E(\mathbf{z}\mathbf{X}^{*\prime})\{\operatorname{diag} D(\mathbf{X}^*)\}^{-\frac{1}{2}}.$$

Now

$$D(\mathbf{z}) = \mathbf{M}^{-1}(\mathbf{M}^{-1})', \quad D(\mathbf{X}^*) = \mathbf{M}'D(\mathbf{X})\mathbf{M} = \mathbf{M}'(\mathbf{\Gamma}^2 + \mathbf{\Gamma})\mathbf{M}$$

and

$$E(\mathbf{zX}^{*\prime}) = E(\mathbf{M}^{-1}\mathbf{yx}'\boldsymbol{\psi}^{-1}\boldsymbol{\Lambda}\mathbf{M}) = E\{\mathbf{M}^{-1}\mathbf{y}(\boldsymbol{\mu} + \boldsymbol{\Lambda}\mathbf{y} + \mathbf{e})'\boldsymbol{\psi}^{-1}\boldsymbol{\Lambda}\mathbf{M}\}$$
$$= \mathbf{M}^{-1}\boldsymbol{\Gamma}\mathbf{M}.$$

We can choose $\mathbf{M}$ so that $\text{diag}\{\mathbf{M}^{-1}(\mathbf{M}^{-1})'\} = \mathbf{I}$ and then

$$\text{corr}(\mathbf{z}, \mathbf{X}^{*\prime}) = \mathbf{M}^{-1}\boldsymbol{\Gamma}\mathbf{M}\{\text{diag}\,\mathbf{M}'(\boldsymbol{\Gamma}^2 + \boldsymbol{\Gamma})\mathbf{M}\}^{-\frac{1}{2}}. \tag{4.28}$$

Similarly,

$$\text{corr}(\mathbf{X}^*) = \{\text{diag}\,D(\mathbf{X}^*)\}^{-\frac{1}{2}}D(\mathbf{X}^*)\{\text{diag}\,D(\mathbf{X}^*)\}^{-\frac{1}{2}}$$
$$= \{\text{diag}\,\mathbf{M}'(\boldsymbol{\Gamma}^2 + \boldsymbol{\Gamma})\mathbf{M}\}^{-\frac{1}{2}}\mathbf{M}'(\boldsymbol{\Gamma}^2 + \boldsymbol{\Gamma})\mathbf{M}\{\text{diag}\,\mathbf{M}'(\boldsymbol{\Gamma}^2 + \boldsymbol{\Gamma})\mathbf{M}\}^{-\frac{1}{2}}. \tag{4.29}$$

For the components to be informative we would want (4.28) to be as near to the unit matrix as possible and (4.29) to be near to corr($\mathbf{z}$). It is not easy to see from the formulae whether this is so, but calculations can easily be made. In the case of the example used throughout this section the estimated correlation between the components of Table 4.3 is 0.721. The data were, in fact, generated artificially from a two-factor model with correlation between the factors of 0.707. In this example, at least, the transformed components correspond very closely to the underlying factors. This supports the general argument of this chapter for using components as proxies for the underlying factors.

The General Case

For the linear model we have shown that there is a close relationship between the components and the underlying factors and, furthermore, that it can be quantified. Similar investigations could be carried out for other models, but we conclude this section with a weaker but very general result which covers all models derived from the sufficiency principle. It says, in effect, that the components provide an ordinal ranking of individuals along the corresponding factor dimension. More precisely we have the

Theorem *If two individuals have* $\mathbf{x}$*-vectors* $\mathbf{x}_1$ *and* $\mathbf{x}_2$ *then*

$$E\{\xi(z_j) \mid \mathbf{x}_1\} \geqslant E\{\xi(z_j) \mid \mathbf{x}_2\}$$

if and only if $X_{j1} \geqslant X_{j2}$ *where* X_{j1} *is the jth component value for the first individual and* X_{j2} *for the second and* ξ *is any monotonic non-decreasing function.* Thus if we rank sample members according to the magnitude of their component value we shall have the same ranking as if we had ranked them according to their expected position on the z-scale, *whatever the distribution of* $\mathbf{z}$.

PROOF

$$E\{\xi(z_j)\mid \mathbf{x}_1\} \geqslant E\{\xi(z_j)\mid \mathbf{x}_2\} \Rightarrow I_S = \int_S \xi(z_j)\{h(\mathbf{z}\mid \mathbf{x}_1) - h(\mathbf{z}\mid \mathbf{x}_2)\}\,\mathrm{d}\mathbf{z} \geqslant 0$$

where S is the sample space of $\mathbf{z}$. Let $\lambda(\mathbf{z}) = h(\mathbf{z}\mid \mathbf{x}_1)/h(\mathbf{z}\mid \mathbf{x}_2)$, A be the set of $\mathbf{z}$'s for which $\lambda \leqslant 1$, B the set for which $\lambda > 1$, $a_j = \sup_A \xi(z_j)$ and $b_j = \inf_B \xi(z_j)$. Then

$$I_S = I_A + I_B$$

where I is the integral of $\xi(z_j)\{h(\mathbf{z}\mid \mathbf{x}_1) - h(\mathbf{z}\mid \mathbf{x}_2)\}$ over the region indicated by the subscript.

Now

$$I_A \geqslant a_j \int_A \{h(\mathbf{z}\mid \mathbf{x}_1) - h(\mathbf{z}\mid \mathbf{x}_2)\}\,\mathrm{d}\mathbf{z}$$

and

$$I_B \geqslant b_j \int_B \{h(\mathbf{z}\mid \mathbf{x}_1) - h(\mathbf{z}\mid \mathbf{x}_2)\}\,\mathrm{d}\mathbf{z}.$$

Since h is a probability density function

$$\int_S \{h(\mathbf{z}\mid \mathbf{x}_1) - h(\mathbf{z}\mid \mathbf{x}_2)\}\,\mathrm{d}\mathbf{z} = 0$$

and, therefore,

$$I_S = I_A + I_B \geqslant (b_j - a_j) \int_B \{h(\mathbf{z}\mid \mathbf{x}_1) - h(\mathbf{z}\mid \mathbf{x}_2)\}\,\mathrm{d}\mathbf{z}.$$

The last integral is positive by the definition of B, and hence $I_S \geqslant 0$ if and only if $b_j \geqslant a_j$. We now show that the necessary and sufficient condition for this to hold is that $\lambda(\mathbf{z})$ is monotonic non-decreasing in z_j. It is obviously sufficient since $\xi(z_j)$ is assumed monotonic non-decreasing in z_j and so every point in B has a value of $\xi(z_j)$ at least as large as any in A. To prove the necessity note that if $\lambda(\mathbf{z})$ were not monotonic non-decreasing in z_j it would be possible, by increasing z_j, to move from a point in B to one in A which had a larger value of $\xi(z_j)$ and this would contradict $b_j \geqslant a_j$.

The conclusion follows from the fact that

$$\lambda(\mathbf{z}) = h(\mathbf{z}\mid \mathbf{x}_1)/h(\mathbf{z}\mid \mathbf{x}_2) = \exp \sum_{j=1}^{q} z_j(X_{j1} - X_{j2})$$

and this is monotonic non-decreasing in z_j if and only if $X_{j1} \geqslant X_{j2}$, as was to be proved.

Factor Scores

Our treatment of the analysis and interpretation of the general factor model does not require a separate discussion of factor scores. By making the components the main vehicle for interpretation and by showing how they are related to the underlying factors we have implied that the components can be used as substitutes for the factors of which they are indicators. In the general case the theorem shows that we are only entitled to infer an ordinal scaling. However, for the linear model, (4.25) holds regardless of the prior distribution of $\mathbf{y}$ and so provides a justification for treating the components in that case as interval level measurements. In this case the components are closely related to the traditional factor scores given in Chapter 3, since if $\mathbf{\Gamma}$ is diagonal the latter are all simply re-scaled components.

4.4 Fitting the models

The discussion so far has proceeded on the assumption that the model parameters were known and virtually all the results were independent of the prior distribution of the factors. Before the methods can be implemented the parameters must be estimated and this will, in general, require an assumption about the distribution. For example, maximum likelihood uses the joint distribution of the x's and this (see (1.1)) depends on $h(\mathbf{y})$. It thus appears that the irrelevance of the prior distribution is illusory and that arbitrariness will re-enter at the estimation stage. However, we shall show in this section that it is often possible to estimate parameters without making distributional assumptions and that even where they are necessary their effect is likely to be small. This conclusion might have been anticipated from our discussion of the linear model in Chapter 3 where least squares methods were used.

The least squares method depended only on second moment properties of the random variables, and we shall now see how other models of the general family can be treated by similar methods.

An assumption will be required about the independence of the latent variables, namely that the z's as defined in (4.4) and appearing in (4.11) are mutually independent. This is a weak assumption since if we wish to admit correlated latent variables a linear transformation of the z's to new variables will leave the linear form of the model unaffected.

Suppose that we can find a transformation $\xi(x_i)$ of x_i such that

$$E\{\xi(x_i) \mid \mathbf{z})\} = \theta_i = \sum_{j=1}^{q} \alpha_{ij} z_j \tag{4.30}$$

then

$$\operatorname{cov}\{\xi(x_i), \xi(x_l)\} = E[E\{\xi(x_i)\xi(x_l) \mid \mathbf{z}\}] - E\xi(x_i)E\xi(x_l), \quad i \neq l$$

$$= \tau^2 \sum_{j=1}^{q} \alpha_{ij}\alpha_{lj} \tag{4.31}$$

and

$$\operatorname{var}\{\xi(x_i)\} = \tau^2 \sum_{j=1}^{q} \alpha_{ij}^2 + E \operatorname{var}\{\xi(x_i) \mid \mathbf{z}\} \tag{4.32}$$

where τ^2 is the common variance of the z's. Thus $D(\xi)$ may be written in the form

$$D(\xi) = \tau^2 \mathbf{A}\mathbf{A}' + \mathbf{V} \tag{4.33}$$

where $\mathbf{V}$ is a diagonal matrix. This is of the same form as (3.4) with $\mathbf{\Lambda} = \boldsymbol{\tau}\mathbf{A}$ and hence the model can be fitted by the least squares methods of Chapter 3. There is no loss of generality in assuming $\tau = 1$ as in the linear factor model.

For this method to be of any use it must be possible to find a transformation ξ with the required property. One non-normal case where this can be done is when

$$g_i(x_i \mid \theta_i) = \theta_i^{\nu_i} x_i^{\nu_i - 1} \exp\{-x_i\theta_i\}/\Gamma(\nu_i) \tag{4.34}$$

where ν_i is a nuisance parameter. It is easy to show that

$$\xi(x_i) = (\nu_i - 1)/x_i$$

has the required property if $\nu_i > 1$. It is unlikely that the ν_i's would be known, but if they were known to be equal an analysis could be based on the reciprocals of the x's since the common factor $(\nu - 1)$ could be absorbed into the α's.

Even if there is no such transformation, as when $\nu_i = 1$ in the foregoing example, it may still be possible to find an approximation. One method of doing so utilizes the fact that for any distribution of exponential type

$$E\{u_i(x_i) \mid \theta_i\} = \frac{-\mathrm{d}}{\mathrm{d}\theta_i} \log G_i(\theta_i) = T(\theta_i), \quad \text{say.}$$

This suggests trying the transformation

$$\xi(x_i) = T^{-1}\{u_i(x_i)\} \tag{4.35}$$

or some modification of it.

As an illustration consider the case when x_i is a Poisson variable with mean m_i. The natural parameter is $\theta_i = \log m_i$, $u_i(x_i) = x_i$ and

$T(\theta_i) = \exp \theta_i$ which suggests the transformation $\xi(x_i) = \log x_i$. However, $\log x_i$ has infinite expectation, but if we take

$$\xi(x_i) = \log(x_i + c)$$

to avoid this we find that with $c = \frac{1}{2}$

$$E \log(x_i + \tfrac{1}{2}) = \theta_i + O(m_i^{-2}). \tag{4.36}$$

Thus, provided that the means are large enough, we could fit the model by least squares using the covariances of the transformed variables.

A similar argument for binomial random variables when x_i has parameters n_i and p_i yields the transformation

$$\xi(x_i) = \log\{(x_i + \tfrac{1}{2})/(n_i - x_i - \tfrac{1}{2})\} \tag{4.37}$$

which requires n_i to be reasonably large. A very important special case which cannot be handled by this means occurs when $n_i = 1$ for all i. This requires the special treatment given in Chapter 6.

Maximum Likelihood

In principle any factor model can be fitted by the method of maximum likelihood. But to do this, assumptions have to be made about the forms of the distributions of the random variables concerned and this, as we have seen, is a somewhat arbitrary matter. It is therefore important to know whether the parameter estimates are likely to be sensitive to any such assumptions we make. Empirical evidence, some of which is reported later, suggests that the choice of prior distribution may not be critical and the following heuristic argument suggests that this is true quite generally. If maximum likelihood is robust in this sense we may continue to choose normal or other convenient priors.

We now show why the choice of h may have little effect on $f(\mathbf{x})$. This can be written

$$f(\mathbf{x}) = \int h(\mathbf{z}) \prod_{i=1}^{p} F_i(x_i) G_i(\theta_i) \exp u_i(x_i)\theta_i \, d\mathbf{z} \tag{4.38}$$

where $\theta_i = \sum_j \alpha_{ij} z_j$. Assuming the z's to be independent and making the probability integral transformation $v_i = H(z_i)$ for all i we find

$$f(\mathbf{x}) = \int \prod_{i=1}^{p} F_i(x_i) G_i(\theta_i) \exp u_i(x_i)\theta_i \, d\mathbf{v}$$

where θ_i is now expressed as a function of $\mathbf{v}$. Finally we make the transformation

$$w_i = \theta_i, \quad (i = 1, 2, \ldots, p)$$

which yields

$$f(\mathbf{x}) = \prod_{i=1}^{p} F_i(x_i) \int \prod_{i=1}^{p} G_i(w_i) \exp\{u_i(x_i) w_i\} f(\mathbf{w}) \, d\mathbf{w} \qquad (4.39)$$

where $f(\mathbf{w})$ is the joint density function of the w's. If q is moderately large, $\mathbf{w}$ will have an approximately multivariate normal distribution because w_i is the sum of q independent and identically distributed random variables. Because this result does not depend on the form of h, $f(\mathbf{x})$ and hence the likelihood will be almost independent of the form of the prior. If q is small, the central limit argument will not apply with the same force and the matter requires further investigation. In the more restricted case where g_i is chosen to be normal and when the model can be written as in (4.9) then, even with $q = 1$, x_i is the sum of two independent random variables one of which is normal. Unless z_1 is highly non-normal the form of the distribution of x_i will not depend critically on that of z_1.

4.5 An extended sufficiency principle

The sufficiency principle as set out in Section 4.2 requires the minimal sufficient statistic to have dimension exactly equal to q. However, there is no reason, apart from a loss of simplicity, for not allowing the dimension to be greater than q. Provided that it is still much smaller than p, the reduction in dimensionality may still be worthwhile, even though the relationship between the factors and the components may then be less direct.

Such an extension is required to accommodate factor models in which the mean of each x_i is assumed to depend on the latent variables in a non-linear (usually polynomial) fashion. Models of this kind have been developed from a traditional point of view by McDonald (1962a, 1967a) and Etezadi-Amoli and McDonald (1983). Their relationship with the approach of this chapter is discussed in Bartholomew (1985a) and Bartholomew and McDonald (1986) (these two papers should be read together). We shall not develop the theory here in a general way but, instead, take a simple one-factor model to illustrate the point at issue.

Suppose that

$$x_i \mid z \frown N(\mu_i + \alpha_i z + \beta_i z^2, \psi_i), \quad (i = 1, 2, \ldots, p) \qquad (4.40)$$

then it is easy to show that the conditional posterior density function of z given $\mathbf{x}$ depends on the x's only through the two components

$$X_1 = \sum_{i=1}^{p} \alpha_i \psi_i^{-\frac{1}{2}} x_i, \quad X_2 = \sum_{i=1}^{p} \beta_i \psi_i^{-\frac{1}{2}} x_i.$$

The dimension of the sufficient statistic is thus 2 and that of the factor space is 1. In a formal sense there is no essential difference between this model and a linear model with two factors. The quadratic term z^2 is treated as if it were a second factor. Since each component is correlated with the corresponding factor we would therefore expect there to be an approximate quadratic functional relationship between X_1 and X_2. The empirical discovery of such a relationship would be indicative of the appropriateness of the quadratic model. In essence this is equivalent to the method used by Etezadi-Amoli and McDonald (1983) though there is, of course, a good deal more to be said about the details of estimation.

4.6 Categorical latent variables

Although we have treated the factors as continuous variables thus far, the general approach based on the sufficiency principle also applies if they are categorical or mixed. Our aim now is to outline the form which the models take and to indicate some of the new practical problems which arise. The simple latent class models already described in Chapter 2 are special cases of the general model but, as we shall show, there is ample scope for new developments.

The essential change is in the specification of the latent variables. A sample member's location on the qth latent dimension is no longer given by the value y of a latent variable but by the label of the latent category into which it falls. Let there be c_j categories numbered $0, 1, \ldots, c_j - 1$ on dimension j, then we may indicate into which category the member falls by a vector-valued indicator variable $\mathbf{u}_j$. If an individual falls into category k, $\boldsymbol{\mu}_j$ will have a 1 in the kth position and zeros elsewhere. The conditional distribution g_i may now be written

$$g_i(x_i \mid \mathbf{u})$$

where $\mathbf{u}' = (\mathbf{u}_1', \mathbf{u}_2', \ldots, \mathbf{u}_q')$ specifies the location of the individual in the latent space. The prior distribution will now be a discrete probability distribution over the $\prod_{j=1}^{q} c_j$ category combinations. If we suppose, as before, that the latent variables exert their influence only on the natural parameter of g_i we shall have

$$\theta_i = \sum_{j=1}^{q} \mathbf{u}_j' \boldsymbol{\alpha}_{ij} \tag{4.41}$$

where $\boldsymbol{\alpha}_{ij}$ is now a vector of dimension c_j. The exponent in

$$\prod_{i=1}^{p} g_i(x_i \mid \mathbf{u})$$

then becomes

$$\sum_{j=1}^{q} \mathbf{u}_j' \sum_{i=1}^{p} \boldsymbol{\alpha}_{ij} x_i = \sum_{j=1}^{q} \mathbf{u}_j' \mathbf{X}_j. \tag{4.42}$$

The component $\mathbf{X}_j = \sum_{i=1}^{p} \boldsymbol{\alpha}_{ij} x_i$ is also a vector of dimension c_j, each of whose elements is associated with the corresponding category on the jth latent dimension. The posterior probability of any latent category combination thus depends on the manifest variables only through the components.

As we have defined it, the model is over-parametrized and the α's are not identifiable. To see this, suppose we increase each element of $\boldsymbol{\alpha}_{ij}$ by a constant amount δ_i to give $\boldsymbol{\alpha}_{ij}^* = \boldsymbol{\alpha}_{ij} + \delta_i \mathbf{1}$, then θ_i becomes

$$\sum_{j=1}^{q} \mathbf{u}_j' \boldsymbol{\alpha}_{ij}^* - \delta_i$$

because $\mathbf{u}_j' \mathbf{1} = 1$, by definition. The constant δ_i can then be absorbed into the other terms in $g_i(x_i \mid \theta_i)$ and nothing is altered. We are thus free to impose constraints to remove the trouble. One possibility is to take $\boldsymbol{\alpha}_{ij}' \mathbf{1} = 0$ and another is to set one element of $\boldsymbol{\alpha}_{ij}$ equal to zero.

In principle, therefore, the extension to categorical latent variables is straightforward, but in practice there are complications especially if $q > 1$. In the first place the number of parameters is liable to be very large. Unless there are prior reasons for fixing the number of categories the c's must be treated as unknowns. If the c's are fixed there are

$$p \sum_{j=1}^{q} (c_i - 1)$$

α's and to this must be added the prior probabilities. If we treat the latent variables as independent there will be $c_j - 1$ unknown prior probabilities for each dimension. Only if $c_j = 2$ for all j is the number of parameters comparable with the case of continuous latent variables.

An important difference is that here the prior probabilities have to be estimated. In the previous case this could be avoided altogether using least squares methods, but even when using maximum likelihood the prior was a specified distribution with no unknown parameters. Here it would, of course, be possible to specify a standard prior distribution but, with an unknown number of categories, it is far from clear how this could be done in a plausible fashion. The essence of the difficulty is that whereas any continuous distribution can be transformed monotonically into any other, the same is not true of two discrete distributions.

An application of a model with two binary latent variables, due to Goodman (1978), was given in Chapter 2. There is no difficulty in principle in introducing more latent variables with more categories but, in practice, the rapid multiplication of parameters to be estimated means that sample sizes would have to be very large indeed if there were to be any prospect of estimating the parameters with sufficient precision to distinguish one model from another. In spite of its generality the model set out here is therefore likely to be limited in its applicability.

With categorical latent variables we cannot compute factor scores as a way of summarizing the posterior probability distribution. We can, however, use this distribution to find the most likely location for an individual in the latent space. Thus if we wish to decide whether an individual with manifest vector $\mathbf{x}$ is more likely to have latent location $\mathbf{u}_1$ or $\mathbf{u}_2$ we compute

$$\frac{h(\mathbf{u}_1 \mid \mathbf{x})}{h(\mathbf{u}_2 \mid \mathbf{x})} = \frac{h(\mathbf{u}_1) \prod_{i=1}^{p} G_i(\mathbf{u}_1) \exp \sum_{j=1}^{q} \mathbf{u}_j'(1)\mathbf{X}_j}{h(\mathbf{u}_2) \prod_{i=1}^{p} G_i(\mathbf{u}_2) \exp \sum_{j=1}^{q} \mathbf{u}_j'(2)\mathbf{X}_j} \tag{4.43}$$

where $\mathbf{u}_j'(l)$ denotes the indicator vector for dimension j, $l = 1, 2$. The relative probability thus depends on the data only through the linear discriminant

$$\sum_{j=1}^{q} \{\mathbf{u}_j'(1) - \mathbf{u}_j'(2)\}\mathbf{X}_j. \tag{4.44}$$

Alternatively, which of two individuals with component values $\mathbf{X}_{j1}$ and $\mathbf{X}_{j2}$ is more likely to fall in location $\mathbf{u}$ is determined by the sign of

$$\sum_{j=1}^{q} \mathbf{u}_j'(\mathbf{X}_{j1} - \mathbf{X}_{j2}). \tag{4.45}$$

We have made no use of any order information about the latent categories. This situation seems unlikely to arise because ordered categories usually imply an underlying metrical variable. In the latent case one would most naturally take the underlying variable itself as the latent variable and use the theory of the main part of the chapter.

The case of mixed-categorical and continuous-latent variables poses no new problems of principle, although it will not be pursued here.

CHAPTER 5

Models for Binary Data

5.1 Preliminaries

An important special case of the general factor model occurs when the x's are binary variables. This situation arises, for example, in educational testing where a child may get an item in a test right or wrong or in a survey where questions are answered yes or no. It is convenient to code the two possible outcomes 1 and 0 so that the data matrix is an $n \times p$ array of zeros and ones. We shall refer to any row of that matrix which gives the set of responses for a given individual as a *response* or *score pattern*. With p variables, each having two outcomes, there are clearly 2^p different response patterns which are possible. If p is small and n is large it is common to summarize the data in a table showing the frequencies of each response pattern (see Tables 9.1, 9.5, 9.8 and 9.10). However, 2^p increases rapidly with p, and a point is quickly reached where 2^p is very much larger than n, so that most of the frequencies are zero or one. In this case it is more economical to list only those response patterns which actually occur. This distinction between large and small p is important when designing the inputs and outputs of computer programs for fitting models and when testing goodness of fit.

Using the notation of previous chapters, the random variable $\mathbf{x}$ is now a p-vector whose elements are 0 or 1 and $f(\mathbf{x})$ is the associated probability function. The relative frequencies of the different response patterns provide estimates of these probabilities. The marginal distributions of $f(\mathbf{x})$ play a key role in the analysis of the models. We shall refer to the univariate distributions $f(x_i)$, $(i = 1, 2, \ldots, p)$ as the first-order margins, the bivariate distributions $f(x_i, x_j)$, $(i \neq j)$ as the second-order margins and so on. The probability of each response pattern can be expressed in terms of marginal probabilities of various orders and this offers practical advantages. In particular, the complete joint probability function of $\mathbf{x}$ is uniquely determined by the following set of marginal probabilities:

$$\left.\begin{array}{l} Pr\{x_i = 1\}, \quad (i = 1, 2, \ldots, p), \\ Pr\{x_i = 1, x_j = 1\}, \quad (i, j = 1, 2, \ldots, p, i \neq j), \\ Pr\{x_i = 1, x_j = 1, x_k = 1\} \quad (i, j, k = 1, 2, \ldots, ;, i \neq j \neq k) \ldots \\ Pr\{x_1 = 1, x_2 = 1, \ldots, x_p = 1\}. \end{array}\right\} \qquad (5.1)$$

This result is easily established by enumerating cases. There are 2^p response patterns, but since the sum of their probabilities is unity, only $2^p - 1$ have to be determined. The number of probabilities in (5.1) is also

$$\binom{p}{1} + \binom{p}{2} + \ldots + \binom{p}{p} = 2^p - 1$$

which implies that the equations connecting the two sets of probabilities have a unique solution for one set in terms of the other.

Two approaches to the construction of models for binary data have been used in the past. One, with its roots in the theory of educational testing and developed further in Bartholomew (1980 and 1981b), starts with a response function giving the probability of a positive response for an individual with latent position $\mathbf{y}$. The other, in the factor analysis tradition, supposes that the binary x's have been produced by dichotomizing underlying continuous variables. As we shall see, the two approaches are equivalent for binary data so that the choice between them is largely a matter of taste. But when we come, in the next chapter, to polytomous responses the two approaches diverge and lead to different methods of analysis. We now deal with each in turn.

5.2 The response function (RF) approach

A general approach to the choice of a suitable response function was set out in Bartholomew (1980). That paper began by listing a set of properties which the family of response functions should possess and then proposed a class of linear models meeting these requirements of the form

$$G^{-1}\{\pi_i(\mathbf{y})\} = \alpha_{i0} + \sum_{j=1}^{q} \alpha_{ij} H^{-1}(y_j), \quad (i = 1, 2, \ldots, p) \qquad (5.2)$$

where $\pi_i(\mathbf{y}) = Pr\{x_i = 1 \mid \mathbf{y}\}$ is the *response function* and y_j ($j = 1, 2, \ldots, q$) are independently and uniformly distributed on $(0, 1)$. The functions G^{-1} and H^{-1} (denoted by G and H in Bartholomew (1980)) were arbitrary but such that their inverses G and H were distribution functions of random variables symmetrically distributed about zero. In educational testing, where $\mathbf{y}$ is usually one-dimensional representing an ability of some kind, $\pi_i(y)$ is referred to as the *item characteristic curve* (ICC) or *item response function* (IRF). The shape of the curve thus shows how the probability of a correct response increases with ability. A similar family of models was arrived at by the general treatment given in Chapter 4. The latter included the requirement that the components be a sufficient summary of the data and this uniquely determines the function G^{-1} as we now show. The essence of the method is to choose the conditional distributions of the x's so that the posterior distribution of $\mathbf{y}$ given $\mathbf{x}$ depends on $\mathbf{x}$ through a q-dimensional vector of functions of $\mathbf{x}$. In the present case this is easily achieved because

$$g_i(x_i \mid \mathbf{y}) = \{\pi_i(\mathbf{y})\}^{x_i}\{1 - \pi_i(\mathbf{y})\}^{1-x_i} \quad (i = 1, 2, \ldots, p) \tag{5.3}$$

which is a member of the exponential family as the theory requires. We thus have

$$h(\mathbf{y} \mid \mathbf{x}) \propto h(\mathbf{y}) \prod_{i=1}^{p} \{\pi_i(\mathbf{y})\}^{x_i}\{1 - \pi_i(\mathbf{y})\}^{1-x_i}$$

$$= h(\mathbf{y}) \prod_{i=1}^{p} \{1 - \pi_i(\mathbf{y})\} \exp \sum_{i=1}^{p} x_i \operatorname{logit} \pi_i(\mathbf{y}). \tag{5.4}$$

Comparison with (4.3) shows that the model needs to be of the form

$$\operatorname{logit} \pi_i(\mathbf{y}) = \alpha_{i0} + \sum_{j=1}^{q} \alpha_{ij} H^{-1}(y_j) \quad (i = 1, 2, \ldots, p). \tag{5.5}$$

This is a special case of (5.2) obtained by choosing G^{-1} to be the logit function. Under these assumptions the components are

$$X_j = \sum_{i=1}^{p} \alpha_{ij} x_i, \quad (j = 1, 2, \ldots, q). \tag{5.6}$$

It was argued in Bartholomew (1980) that H should be the distribution function of a symmetrically distributed random variable, but this still leaves us with a wide choice. However, the two functions which have been used most often in practice are the logistic and the normal. These are very similar in shape, and the choice between them is therefore of little practical importance. There is, nevertheless, a good reason for choosing the normal function if there are two or more latent variables as we now show.

We recall that in the "normal" factor model for continuous variables the axes of the factor space may be rotated without affecting the fit of the model. We saw (in 3.3) that this could be expressed either as a transformation of the latent variables or of the loadings. The possibility of making such transformations in the normal case allows one to choose from among the set of solutions that which has most substantive meaning. One would expect there to be a parallel result for binary data and we now investigate whether this is so.

We first make the transformation $z_j = H^{-1}(y_j)$, $(j = 1, 2, \ldots, q)$ which gives

$$f(\mathbf{x}) = \int_{-\infty}^{+\infty} \cdots \int \prod_{i=1}^{p} \{\pi_i(\mathbf{z})\}^{x_i}\{1 - \pi_i(\mathbf{z})\}^{1-x_i} h(\mathbf{z})\, d\mathbf{z} \tag{5.7}$$

where $\pi_i(\mathbf{z}) = G\left(\alpha_{i0} + \sum_{j=1}^{q} \alpha_{ij} z_j\right)$ and h is the density function associated with H. Let $\mathbf{A} = \{\alpha_{ij}\}$, $(i = 1, 2, \ldots, p; j = 1, 2, \ldots, q)$, and let $\mathbf{A}_i$ denote the ith row of $\mathbf{A}$. If we now introduce new parameters defined by

$$\mathbf{A}^* = \mathbf{AM}$$

where $\mathbf{M}$ is a $q \times q$ orthogonal matrix, then $\sum_{j=1}^{q} \alpha_{ij} z_j = \mathbf{A}_i \mathbf{z} = (\mathbf{A}^*\mathbf{M}^{-1})_i \mathbf{z}$. Finally we set $\mathbf{z}^* = \mathbf{M}^{-1}\mathbf{z}$. Since $\mathbf{M}$ is orthogonal the Jacobian of the transformation is 1 and

$$\pi_i(\mathbf{z}^*) = G(\alpha_{i0} + \mathbf{A}_i^* \mathbf{z}^*).$$

Referring back to (5.7), it appears that $f(\mathbf{x})$ will be unchanged by the transformation if the joint distribution of $\mathbf{z}^*$ is the same as that of $\mathbf{z}$. It was shown by Lancaster (1954) that if both $\mathbf{z}$ and $\mathbf{z}^*$ are to be independent under orthogonal rotation then they must also be normal.

We thus arrive at a unique model which may be written in terms of uniformly distributed latent variables $\{y_i\}$ as

$$\text{logit}\, \pi_i(\mathbf{y}) = \alpha_{i0} + \sum_{j=1}^{q} \alpha_{ij} \Phi^{-1}(y_j) \tag{5.8a}$$

or, in terms of normally distributed variables $\{z_j\}$, as

$$\text{logit}\, \pi_i(\mathbf{z}) = \alpha_{i0} + \sum_{j=1}^{q} \alpha_{ij} z_j. \tag{5.8b}$$

We shall call this the *logit/probit model* or, more briefly, the *logit model.* It may be noted that an orthogonal transformation of the weights, $\mathbf{A}$, is equivalent to an orthogonal transformation in the space of $\mathbf{z}$ but not of $\mathbf{y}$.

An important special case of the model arises when $q = 1$ and $\alpha_{i1} = \alpha$ for all i. The components are then proportional to $\sum x_i$ and so they can be determined, apart from a scale factor, without need to estimate the parameters. This version of the model is essentially the same as the Rasch model (Rasch (1960)) from which it differs only that in the latter each individual's position on the latent scale is treated as a fixed parameter instead of a random variable.

Key Properties

Although it is implicit in the general theory, it may be helpful to draw attention to the two key properties which response functions produced by this approach possess.

The choice of which of the two possible outcomes is to be regarded as "positive" is entirely arbitrary. It ought not to matter, for example, whether the code 1 is attached to "Yes" or "No". If "Yes" has probability $\pi_i(\mathbf{z})$ and "no" has probability $1-\pi_i(\mathbf{z})$ then both $\pi_i(\mathbf{z})$ and its complement should have the same form. In the present case,

$$\pi_i(\mathbf{z}) = \left\{1 + \exp\left(\alpha_{i0} + \sum_{j=1}^{q} \alpha_{ij} z_j\right)\right\}^{-1} \quad \text{and}$$

$$1 - \pi_i(\mathbf{z}) = \left\{1 + \exp\left(-\alpha_{i0} - \sum_{j=1}^{q} \alpha_{ij} z_j\right)\right\}^{-1}.$$

Both have the same form, the only difference being that the signs of all the α's are changed. This is as it should be since if increasing any z increases the probability of answering "Yes" it should obviously decrease the probability of answering "No" by the same amount. This property is useful in estimation. For example, with $q=1$ it is possible to ensure that all α_{i1}'s are positive or zero by an appropriate choice of which outcome is to be regarded as positive. This is advantageous with most estimation methods.

The second property stems from the arbitrariness of the direction in which most latent variables are measured. Thus if we imagine the political spectrum extending from extreme left to extreme right, the choice of which end of our scale is "left" and which is "right" is of no significance. Changing the direction of measurement involves replacing y_j by $(1-y_j)$ in (5.8a) or z_j by $-z_j$ in (5.8b). This is equivalent to changing the sign of α_{ij} (not *all* of the α's as in the previous case). Again this does not change the form of the model.

Interpretation of the Parameters

The parameters of the logit model may be interpreted in a variety of ways. The coefficient α_{i0} is the value of logit $\pi_i(\mathbf{z})$ at $\mathbf{z}=\mathbf{0}$. It thus defines the probability of a positive response for the median individual. A rather more useful interpretation arises from transforming to new parameters defined by $\alpha_{i0} = \text{logit}\, \pi_i$ $(i = 1, 2, \ldots, p)$; π_i is then the probability of a positive response from the "median" sample member.

The α_{ij}'s have three related interpretations. In the first, α_{ij} is a measure of the extent to which the variable z_j discriminates between individuals. For two individuals a given distance apart on the z_j-scale, the bigger the absolute value of α_{ij} the greater the difference in their probabilities of giving a positive response and hence the easier to discriminate between them on the evidence of their responses to item i. In educational testing, where $q=1$, this is the interpretation usually

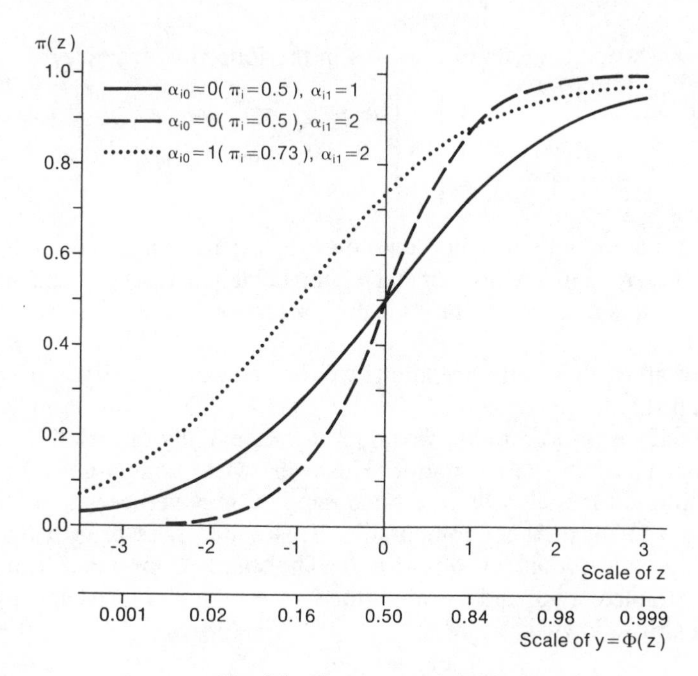

Fig. 5.1 The response function $\pi(z)$

adopted. In the same context α_{i0} or π_i would be termed "difficulty" parameters since an increase in either increases the chance of a correct response (actually they measure the "easiness" of the item). Fig. 5.1 illustrates the effect of differing values of α_{i0} and α_{i1} on the shape of the response curve.

A second interpretation of the α_{ij}'s is by analogy with principal components analysis or linear factor analysis where the α_{ij}'s are regarded as loadings or weights. They are the coefficients of the components as shown in (5.6) and thus indicate the weight to be attached to the various x_i's in the interpretation of the component X_j.

A third and important interpretation of the α_{ij}'s is as category scores. This interpretation is the key to the connection with correspondence analysis to which we shall come later. Suppose that we attach the score α_{ij} instead of 1 to a positive response on variable i and zero to a negative response (there is a different score for each latent variable). Then X_j would be the score obtained by adding up the response scores corresponding to latent variable j for any individual. The *individual score* is thus the sum of the *category scores* for that individual. For the first latent variable, then, we could imagine the data matrix of binary variables being replaced by a new *score pattern matrix* whose elements were α's as in the following example:

$$\begin{bmatrix} 1 & 0 & 1 & 1 & 1 & 0 & 1 \\ 0 & 0 & 1 & 1 & 0 & 1 & 1 \\ & & \vdots & & \vdots & & \end{bmatrix} \rightarrow \begin{bmatrix} \alpha_{11} & 0 & \alpha_{31} & \alpha_{41} & \alpha_{51} & 0 & \alpha_{71} \\ 0 & 0 & \alpha_{31} & \alpha_{41} & 0 & \alpha_{61} & \alpha_{71} \\ & & & \vdots & \vdots & & \end{bmatrix}$$

Instead of adding up the rows of the original matrix of indicator variables to get a score for each individual (as teachers and others often do) we add up the category scores given in the right-hand matrix.

Not all of these interpretations will be relevant in every application but all should be considered.

Whatever interpretation we adopt it is essentially the relative values of the α_{ij}'s ($i \geqslant 1$) that matter. However, when comparing estimates for different models from the same family it may be necessary to scale the α's to make them comparable. This need arises principally with the logit and probit functions for H. The standard logistic distribution has variance $\pi^2/3$ and because this is very similar in shape to the normal it follows that

$$\text{logit}\, u \doteqdot \frac{\pi}{\sqrt{3}} \Phi^{-1}(u).$$

Hence the model

$$\text{logit}\, \pi_i(y) = \alpha_{i0} + \sum_{j=1}^{q} \alpha_{ij} \Phi^{-1}(y)$$

is approximately the same as

$$\text{logit}\, \pi_i(y) = \alpha_{i0} + \sum_{j=1}^{q} \alpha_{ij} (\sqrt{3}/\pi)\, \text{logit}\, y_j.$$

Thus if we fit the logit/logit model we would expect the weights to be $(\sqrt{3}/\pi)$ times what we would have got with the logit/probit model. Similarly the probit/probit model

$$\Phi^{-1}(\pi_i(y)) = \alpha_{i0} + \sum_{j=1}^{q} \alpha_{ij} \Phi^{-1}(y_j)$$

is approximately the same as the logit/logit model

$$\text{logit}\, \pi_i(y) = \alpha_{i0}\pi/\sqrt{3} + \sum_{j=1}^{q} \alpha_{ij}\, \text{logit}\, y_j.$$

In this case the weights are unaffected but the α_{i0}'s have to be adjusted as shown. We shall take the logit/probit model as the standard and estimates from other models will be scaled as necessary.

This point should be borne in mind when comparing estimates quoted here with those already published relating to these models.

Factor Scores

The problem of finding factor scores is not essentially different to what it was with the normal model. The components are linear combinations of the x's and we know from the general theory that the expectation of any monotonic function of y_j will order the individuals in the same way as the component X_j. For many purposes we shall need to go no further than this. However, there is no objection to using some function of X_j such as is provided by either of the posterior expectations $E(\mathbf{z} \mid \mathbf{x})$ and $E(\mathbf{y} \mid \mathbf{x})$. These can be computed by numerical integration of the posterior distribution if required. There is a theoretical and a practical advantage in using $E(\mathbf{y} \mid \mathbf{x})$. On the theoretical side, because y_j is uniform, it may be interpreted as the proportion of the population lying below that value on the jth latent dimension. Its conditional expectation, given $\mathbf{x}$, thus tells us the expected proportion of the population lying below an individual with that value of $\mathbf{x}$. Given that there is no natural metric for scaling individuals, it is convenient to express their location on the latent dimension by reference to the quantile of the prior distribution at which we would expect to find them. The practical advantage is that when $q = 1$, $E(y \mid \mathbf{x})$ is approximately a linear function of X and hence the numerical integration may be avoided. In this case,

$$E(y \mid \mathbf{x}) = \int_0^1 y^{X+1}(1-y)^{A-X}\psi(y)\,\mathrm{d}y \Big/ \int_0^1 y^X(1-y)^{A-X}\psi(y)\,\mathrm{d}y \quad (5.9)$$

where $A = \sum_{i=1}^{p} \alpha_{i1}$, $X = \sum_{i=1}^{p} \alpha_{i1} x_i$ and

$$\psi(y) = \left[\prod_{i=1}^{p} \{\pi_i y^{\alpha_{i1}} + (1-\pi_i)(1-y)^{\alpha_{i1}}\}\right]^{-1}.$$

If the α's are small, a linear approximation may be obtained by a Taylor expansion, but if $\psi(y)$ is slowly varying over the range of y for which the remainder of the integral is appreciable we may approximate (5.9) by treating it as a constant, when we obtain

$$E(y \mid \mathbf{x}) = (1+X)/(2+A). \quad (5.10)$$

This result is exact if $\pi_i = \frac{1}{2}$ and $\alpha_{i1} = 1$ for all i, so we may expect the approximation to be best in the neighbourhood of this point. Even when the agreement is not good, calculations in Bartholomew (1984) suggest that the relationship is approximately linear. In any event

(5.10) will always provide the same ranking of individuals as X or either of the conditional expectations.

Other Models

Although we have aimed to show that the logit model is a natural choice for the factor analysis of binary data this is not to say that there are no circumstances in which other models would be preferable. For example, in educational testing where the items to be answered are of the multiple choice kind, a candidate at the extreme lower end of the ability range may give a correct response by guessing. This can be accommodated by allowing $\pi_i(y)$ to approach an asymptote at the lower end greater than zero. Similarly, a person at the opposite extreme may get the item wrong by carelessness or forgetting. However, this modification means that logit $\pi_i(y)$ no longer has the required form.

Another important case which does not meet all the requirements is the two-category *latent class model* discussed in Chapter 2 in which individuals are supposed to belong to one of two latent classes. Suppose those in the lower class respond positively on variable i with probability π_{0i} and those in the upper class with probability π_{1i}. We can then think of this as a special case of (5.2) with $q = 1$ and with G^{-1}

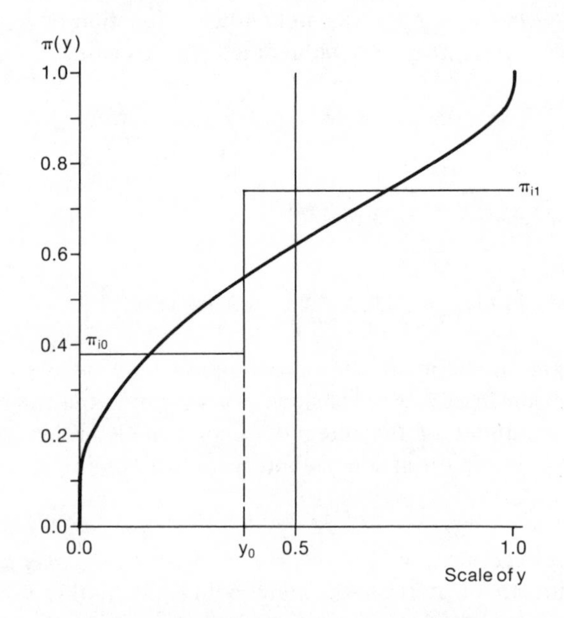

Fig. 5.2 A response function for a latent class model (step function) compared with the logit response function with $\pi_i = 0.62$ and $\alpha_{i1} = 1.0$

as a step function. We define a critical value y_0 such that $\pi_i(y) = \pi_{i0}$ for $y \leqslant y_0$ and $\pi_i(y) = \pi_{i1}$ for $y > y_0$. The position is illustrated in Fig. 5.2. This model is not covered by the general theory based on the exponential family because here G^{-1} is not the logistic function, but it may easily be shown that the posterior probability of being in either class depends on the x's through a single linear function of the x's (see Aitkin (1980)).

There are other models which do not belong to the family we have defined but which differ from it by so little that the difference can be ignored. One such arises in the next section where we shall meet arguments for choosing $G = \Phi$, the normal distribution function, which leads to the probit/probit model.

5.3 The underlying variable (UV) approach

The approach to binary data which has been used in the factor analysis tradition appears at first sight to be rather different. Its aim has been to bring the analysis within the scope of the traditional linear model for the factor analysis of manifest variables. This is achieved by supposing that underlying the ith dichotomy there is a continuous variable, ξ_i say. The observed binary variable x_i is then an indicator of whether ξ_i is above or below some critical level τ_i. Thus we may define

$$\begin{aligned} x_i &= 1 \quad \text{if} \quad \xi_i \leqslant \tau_i \\ &= 0 \end{aligned} \tag{5.11}$$

where

$$\boldsymbol{\xi} = \boldsymbol{\mu} + \boldsymbol{\Lambda}\mathbf{z} + \mathbf{e}. \tag{5.12}$$

If the usual assumptions are made about the distributions of $\mathbf{z}$ and $\mathbf{e}$ then the model could be fitted if we had estimates of the covariance or correlation matrix of $\boldsymbol{\xi}$. If $\mathbf{z}$ and $\mathbf{e}$ are both assumed to be normal then $\boldsymbol{\xi}$ will be normal and there are then well-known methods of estimating the correlation coefficients from the bivariate 2×2 tables of the x's. These yield the so-called tetrachoric correlations which can be used as input to a standard factor analysis program. Unfortunately, correlation matrices estimated by these means are not always positive definite, but more efficient methods based on generalized least squares are available (Muthén (1978)) and will be discussed in Chapter 6.

The choice of a normal distribution for $\mathbf{z}$ and $\mathbf{e}$ and hence for $\boldsymbol{\xi}$ is prompted by the desire to make the model consistent with the standard theory for continuous variables. It is perfectly possible to assume other forms, and there are practical advantages in supposing that the bivariate continuous distribution underlying each of the 2×2

tables is what Mardia (1970) has called a C-type distribution. Such a distribution has the property that, wherever the dichotomy on each variable is made, the cross-product ratio (odds ratio) is the same. Since the threshold values may well have been determined quite arbitrarily, it is reasonable to insist that the results of the analysis should not depend on where the cut was made in any instance. A further advantage of this assumption is that, for some members of the family at least, the estimation of the correlation coefficients is much easier than for the tetrachoric coefficients. Against this must be set the fact that a C-type distribution for (ξ_i, ξ_j) is not consistent with a linear model like (5.12) since bivariate C-type distributions do not arise as linear combinations of random variables. Other possible distributions for $\boldsymbol{\xi}$ could be derived from (5.12) by giving $\mathbf{z}$ and/or $\mathbf{e}$ non-normal distributions, but the central limit effect of forming linear combinations by (5.12) would suggest that $\boldsymbol{\xi}$ would always be approximately normal. It therefore seems preferable to regard methods based on C-type distributions as convenient approximations to the normal model and hence we shall return to them in that connection in Chapter 6.

There is a further matter to be raised about the UV model. Is it always reasonable to postulate an underlying variable? In some cases it undoubtedly is. Indeed the dichotomies may have been arrived at in just that way as when, for example, incomes are recorded as being above or below a tax threshold. Again, if a person is asked whether or not they are in favour of some proposition they may be imagined to have a quantifiable degree of enthusiasm which is imperfectly revealed by their response to a question of the yes/no variety. But there are other cases where the notion is less appealing. The question asked may be a simple matter of fact about which it is difficult to conceive of anything between a categorical yes or no—for example, "Were you born in this country?". However, these considerations create no practical difficulties since, as we now show, the RF and UV models are equivalent for binary data and so the analysis does not depend on the choice of this particular interpretation. The same is not true of polytomous data which we discuss in Chapter 7.

5.4 The equivalence of the RF and UV approaches

To show that the two approaches are equivalent we must show that the joint probability distributions of $\mathbf{x}$ to which they lead are the same. For if this is so then no empirical information can distinguish between them. Instead of working with the probability function $f(\mathbf{x})$ we use the equivalent set of marginal probabilities $\{Pr\{x_i = 1, x_j = 1, \ldots, \}\}$ defined by (5.1). Let S denote any non-empty sub-set of $\{1, 2, \ldots, p\}$, then

$$Pr\left\{\bigcap_{i\in S}(x_i=1)\right\}=\int_{-\infty}^{+\infty}\dots\int Pr\left\{\bigcap_{i\in S}(x_i=1)\mid \mathbf{z}\right\}h(\mathbf{z})\,d\mathbf{z}$$
$$=\int_{-\infty}^{+\infty}\dots\int \prod_{i\in S} Pr\{x_i=1\mid \mathbf{z}\}h(\mathbf{z})\,d\mathbf{z} \qquad (5.13)$$

because of the conditional independence of the x's. In the RF approach $Pr\{x_i=1\mid \mathbf{z}\}$ is the response function and $h(\mathbf{z})$ is the prior distribution of $\mathbf{z}$. To show the equivalence with the UV approach we have to show in this case that the left-hand probability of (5.13) admits the same representation. Thus,

$$Pr\left\{\bigcap_{i\in S}(x_i=1)\right\}=Pr\left\{\bigcap_{i\in S}(\xi_i\leqslant\tau_i)\right\}$$
$$=\int_{-\infty}^{+\infty}\dots\int Pr\left\{\bigcap_{i\in S}(\xi\leqslant\tau_i)\mid \mathbf{z}\right\}h(\mathbf{z})\,d\mathbf{z}. \qquad (5.14)$$

When $\mathbf{z}$ is fixed the ξ's are independent (because the e's are independent) and therefore

$$Pr\left\{\bigcap_{i\in S}(x_i=1)\right\}=\int_{-\infty}^{+\infty}\dots\int \prod_{i\in S} Pr\{\xi_i\leqslant\tau_i\mid \mathbf{z}\}h(\mathbf{z})\,d\mathbf{z}. \qquad (5.15)$$

Now

$$Pr\{\xi_i\leqslant\tau_i\mid z\}=Pr\left\{\mu_i+e_i+\sum_{j=1}^{q}\lambda_{ij}z_j\leqslant\tau_i\mid \mathbf{z}\right\}$$
$$=Pr\left\{\frac{e_i}{\psi_i^{\frac{1}{2}}}\leqslant\frac{\tau_i-\mu_i-\sum_{j=1}^{q}\lambda_{ij}z_j}{\psi_i^{\frac{1}{2}}}\,\middle|\,\mathbf{z}\right\}$$
$$=R\left\{\frac{\tau_i-\mu_i-\sum_{j=1}^{q}\lambda_{ij}z_j}{\psi_i^{\frac{1}{2}}}\right\} \qquad (5.16)$$

where $R(.)$ is the distribution function of $e_i/\psi_i^{\frac{1}{2}}$ (note that in the factor model this does not depend on i). For the RF model,

$$Pr\{x_i=1\mid \mathbf{z}\}=G\left(\alpha_{i0}+\sum_{j=1}^{q}\alpha_{ij}z_j\right). \qquad (5.17)$$

Equivalence between the two models therefore exists if

$G\equiv R$ and if $\alpha_{i0}=(\tau_i-\mu_i)/\psi_i^{\frac{1}{2}}$ and

$$\alpha_{ij}=-\lambda_{ij}/\psi_i^{\frac{1}{2}}\quad(i=1,2,\dots,p). \qquad (5.18)$$

Thus for any model of the one kind we can find a model of the other which is empirically indistinguishable from it. For example, if G^{-1} is the logit function,

$$G(u) = (1 + e^{-u})^{-1} = R(e_i/\psi_i^{\frac{1}{2}}).$$

The standard logistic random variable with distribution function $G(u)$ has mean zero and variance $\pi^2/3$; e_i therefore has a logistic distribution with mean zero and variance $\psi_i\pi^2/3$.

The equivalence of the two methods has been noted by, for example, Tanaka and de Leeuw and by Moran but does not appear to have appeared in print.

An important corollary of the foregoing analysis is that the parameters $\{\tau_i\}$, $\{\mu_i\}$ and $\{\lambda_{ij}\}$ of the UV model are not individually estimable. The likelihood is a function of the parameters $\{(\tau_i - \mu_i)/\psi_i^{\frac{1}{2}}\}$. This is a reflection of the fact that there can be no information about the standard deviation of the underlying ξ's in the dichotomized data. In practice, when working with continuous variables, we usually throw this information away by first standardizing the x's; that is, we assume that

$$\text{var}(\xi_i) = 1 = \sum_{j=1}^{q} \lambda_{ij}^2 + \psi_i.$$

The loadings that we estimate are then

$$\lambda_{ij} \Big/ \left(1 - \sum_j \lambda_{ij}^2\right)^{\frac{1}{2}} = \lambda_{ij}/\psi_i^{\frac{1}{2}} \quad (i, j = 1, 2, \ldots, p; i \neq j)$$

(see Section 4.3 where asterisks were used to distinguish the scale-invariant parameters). For later reference we identify that member of the UV family which is equivalent to the logit model. In that case G^{-1} is the logit function, so

$$G(u) = \frac{1}{1 + e^{-u}}.$$

This is the distribution function of a symmetrically distributed random variable with zero mean and variance $\pi^2/3$. In the UV model $e_i/\psi_i^{\frac{1}{2}}$ thus has to have a distribution of this form which implies that e_i has a logistic distribution with zero mean and variance $\psi_i\pi^2/3$ for all i.

CHAPTER 6

Methods for Binary Data

6.1 Efficient methods of fitting

Methods for fitting models to binary data are more recent and less thoroughly explored than those for continuous data. Nevertheless there is a wide choice of methods available. In this section we discuss efficient methods based on maximum likelihood and generalized least squares. We shall concentrate on the principles underlying the methods rather than on the technical details for which the reader is referred to original sources.

The direct approach to maximizing the likelihood function using standard numerical optimization methods is possible, though even with large computers it is effectively limited to the case of one latent variable and a small (say, less than 10) number of manifest variables. This method is useful, however, as a check on the accuracy of approximate methods even when it is too time-consuming for general use.

For any model of the family (5.2) the joint probability function is

$$f(\mathbf{x}) = \int_{-\infty}^{+\infty} \cdots \int \prod_{i=1}^{p} \{\pi_i(\mathbf{z})\}^{x_i}\{1 - \pi_i(\mathbf{z})\}^{1-x_i} h(\mathbf{z})\, \mathrm{d}\mathbf{z}. \tag{6.1}$$

Let $\mathbf{x}_h$ be the response vector observed for the hth sample member, then the log-likelihood is

$$L = \sum_{h=1}^{n} \log f(\mathbf{x}_h). \tag{6.2}$$

Bock and Lieberman (1970) maximized this function with respect to the model parameters when G and H were both standard normal distribution functions. They used the method of scoring but found that the heavy numerical integration involved meant that the method was limited to one latent variable and 5 or so manifest variables. With larger and faster computers the scope of the method has not been significantly increased.

A much more successful approach, pioneered in this connection by Bock and Aitkin (1981), is based on the E-M algorithm. We shall describe two versions of this method as it applies to the logit model with one latent variable. The extension to several latent variables is straightforward in principle but there are still technical problems to be solved.

An E-M Algorithm

We shall use the version of the model expressed in terms of the normally distributed z-variable given in (5.8b). The method breaks down the operation into an E(expectation)-stage and an M(maximization)-stage which are repeated alternately until convergence is attained. The first variant depends on the following two facts:

(i) If the values of z were known for each individual, the problem would be one of regression with a binary dependent variable as in probit analysis. Fitting such a model by maximum likelihood is straightforward and quite large problems can be speedily solved.

(ii) If the values of the α's were known, the values of z for each individual could be predicted from the posterior distribution given $\mathbf{x}$.

The algorithm for estimating the parameters then proceeds as follows:

Step 1 Assume arbitrary starting values for the parameters.

Step 2 Using these values, predict z for each individual using the posterior expectation of z given $\mathbf{x}$.

Step 3 Treating these expected values as if they were true values, estimate the parameters by maximum likelihood.

Step 4 Return to Step 2 and repeat the cycle until convergence is attained.

Step 3, the M-step, depends on the following analysis. If z_h is the currently predicted value of z for the hth individual then the conditional log-likelihood function is

$$L_z = \sum_{h=1}^{n} \sum_{i=1}^{p} [x_{ih} \log \pi_i(z_h) + \{1 - x_{ih}\} \log\{1 - \pi_i(z_h)\}]$$

$$= \sum_{h=1}^{n} \sum_{i=1}^{p} [x_{ih} \operatorname{logit} \pi_i(z_h) + \log\{1 - \pi_i(z_h)\}]. \tag{6.3}$$

Substituting for $\pi_i(z)$ from (5.8b) then gives

$$L_z = \sum_{h=1}^{n} \sum_{i=1}^{p} [x_{ih}\{\alpha_{i0} + \alpha_{i1} z_h\} - \log\{1 + \exp(\alpha_{i0} + \alpha_{i1} z_h)\}$$

$$= \sum_{i=1}^{p} \alpha_{i0} \sum_{h=1}^{n} x_{ih} + \sum_{i=1}^{p} \alpha_{i1} \sum_{h=1}^{n} x_{ih} z_h$$

$$- \sum_{i=1}^{p} \sum_{h=1}^{n} \log\{1 + \exp(\alpha_{i0} + \alpha_{i1} z_h)\}. \tag{6.4}$$

The partial derivatives with respect to the parameters $\{\alpha_{i0}\}$ and $\{\alpha_{i1}\}$ are then

$$\left.\begin{aligned}\frac{\partial L}{\partial \alpha_{i0}} &= \sum_{h=1}^{n} x_{ih} - \sum_{h=1}^{n} \{1+\exp(-\alpha_{i0}-\alpha_{i1}z_h)\}^{-1} \\ &= \sum_{h=1}^{n} \{x_{ih} - \pi_i(z_h)\} \\ \frac{\partial L}{\partial \alpha_{i1}} &= \sum_{h=1}^{n} x_{ih}z_h - \sum_{h=1}^{n} z_h\{1+\exp(-\alpha_{i0}-\alpha_{i1}z_h)\}^{-1} \\ &= \sum_{h=1}^{n} z_h\{x_{ih} - \pi_i(z_h)\} \qquad (i=1,2,\ldots,p).\end{aligned}\right\} \qquad (6.5)$$

Setting the partial derivatives equal to zero then provides the estimating equations. For each value of i there is a pair of equations for α_{i0} and α_{i1}. The basic step is thus that of solving two non-linear equations in two unknowns. Methods of solving these equations are briefly reviewed by McFadden (1982).

An interesting point to notice about (6.5) is that each pair of equations equates a simple sample function to its expectation. Thus $E\sum_{h=1}^{n} x_{ih} = \sum_{h=1}^{n} \pi_i(z_h)$ and similarly for the second member of the pair. This is a consequence of our choosing a member of the exponential family for the conditional response distribution. The probit/probit model, for example, does not lead to such simple equations.

The E-step requires the computation of $e(z \mid \mathbf{x}_h)$ for each h given the current estimates of $\{\alpha_{i0}\}$ and $\{\alpha_{i1}\}$. These have to be found by numerical integration from

$$E(z \mid \mathbf{x}) = \int_{-\infty}^{+\infty} z \prod_{i=1}^{p} \{\pi_i(z)\}^{x_i}\{1-\pi_i(z)\}^{1-x_i} h(z)\,\mathrm{d}z / f(\mathbf{x}). \qquad (6.6)$$

The value of $\{E(z \mid \mathbf{x}_h)\}$ at the final iteration gives the z-scores.

More than One Latent Variable

If there is more than one latent variable the formulae in (6.3), (6.4) and (6.5) require only minor modifications. In the first part of (6.3) $\alpha_{i1}z_h$ is replaced by $\sum_{j=1}^{q} \alpha_{ij}z_{jh}$. The form of the partial derivatives in (6.5) is the same, but q equations replace the second member—one for each $\{\alpha_{ij}\}$; $\pi_i(z_h)$ now becomes $\pi_i(\mathbf{z}_h)$ and the determination of $\hat{\alpha}_{i0}$ and $\hat{\alpha}_{i1}, \hat{\alpha}_{i2} \ldots \hat{\alpha}_{ij}$ now involves the solution of $q+1$ simultaneous non-linear equations for each i.

The new and important feature which arises when $q > 1$ is that there is no unique solution because of the fact that orthogonal transformations of the α_{ij}'s ($j > 0$) leave the value of the likelihood unchanged. In order to obtain a unique solution we must therefore impose some constraints. If we take the underlying variable viewpoint it is natural to impose the same constraint as we would use if the ξ's were directly observed, namely that $\mathbf{\Lambda}'\boldsymbol{\psi}^{-1}\mathbf{\Lambda}$ be diagonal. As we see from (5.18) this is equivalent to having $\mathbf{A}'\mathbf{A}$ diagonal where $\mathbf{A} = \{\alpha_{ij}\}$. This is easily done if we use the approximate method involving estimated correlation coefficients described below, but is less straightforward if we try to incorporate it into the E-M algorithm. At the maximization stage we would have to include the constraints on the α's, and although straightforward in principle the idea has yet to be implemented.

Another and simpler possibility is to fix the values of enough α's to ensure a unique solution. For example, if $q = 2$, it is sufficient to set $\alpha_{i2} = 0$ for some i. Since, given any solution, we can always find an orthogonal transformation which will make any chosen α_{ij} ($j > 0$) zero, we lose no generality by placing such a restriction on the solution. Yet another possibility is first to fit a one-factor model and then to estimate the α's for the second factor, treating the first factor loadings as given, but this will not, in general, correspond to a maximum of the likelihood function. Additional factors can then be added in the same fashion.

In the current state of knowledge and given the accuracy of the approximation to be discussed later, it seems preferable to use that approximation if $q \geqslant 2$. If necessary it could then be used as a starting point for the iterative solution of the full likelihood equations. It should be noted, however, that the condition that $\mathbf{\Lambda}'\boldsymbol{\psi}^{-1}\mathbf{\Lambda}$ be diagonal does not imply that the components are uncorrelated, but it would ensure that the components constructed from the ξ's would be orthogonal (but not those from the x's). The components given by $\mathbf{X} = \mathbf{A}'\mathbf{x}$ have dispersion matrix

$$D(\mathbf{X}) = \mathbf{A}'D(\mathbf{x})\mathbf{A}$$

where $D(\mathbf{x}) = \{E\pi_i(\mathbf{z})\pi_j(\mathbf{z}) - E\pi_i(\mathbf{z})E\pi_j(\mathbf{z})\}$. This cannot be expressed in simple form, though if the α's are small we can utilize the Taylor series approximation to give

$$D(\mathbf{x}) \doteqdot \left\{\pi_i(1-\pi_i)\pi_j(1-\pi_j)\sum_{k=1}^{q} \alpha_{ik}\alpha_{jk}\right\}.$$

$D(\mathbf{X})$ may then be written

$$D(\mathbf{X}) \doteqdot \{\mathbf{A}'\mathbf{\Pi}(\mathbf{I}-\mathbf{\Pi})\mathbf{A}\}^2 \qquad (6.7)$$

where Π is diagonal with π_i in the ith diagonal position. Thus, even in the limited range of circumstances in which the approximation is justified, the components would only be uncorrelated if the π_i's were all equal.

A Variation on the E-M *Algorithm*

As noted above, the simple form of the equations in the estimation part of the algorithm depends on choosing G^{-1} to be the logit function. However, the same idea can be used for other members of the general family and it was for the probit/probit version that Bock and Aitkin (1981) proposed a variation of the method we have discussed. We shall set it in a rather different perspective in order to demonstrate both its versatility and its limitations.

The method may be viewed as an adaptation of the algorithm already given for the case when z, the latent variable, has a discrete distribution. Thus suppose that z takes the values $z_1, z_2, \ldots, z_k$ with probabilities $h(z_1), h(z_2), \ldots, h(z_k)$. Then

$$f(\mathbf{x}_h) = \sum_{t=1}^{k} f(\mathbf{x}_h \mid z_t) h(z_t) \quad (h = 1, 2, \ldots, n) \tag{6.8}$$

where

$$f(\mathbf{x}_h \mid z_t) = \prod_{i=1}^{p} \{\pi_i(z_t)\}^{x_{ih}} \{1 - \pi_i(z_t)\}^{1-x_{ih}}.$$

We then have to maximize

$$L = \sum_{h=1}^{n} \log f(\mathbf{x}_h)$$

and

$$\begin{aligned} \frac{\partial L}{\partial \alpha_{il}} &= \sum_{h=1}^{n} \frac{\partial f(\mathbf{x}_h)}{\partial \alpha_{il}} \bigg/ f(\mathbf{x}_h) \quad (l = 0, 1) \\ &= \sum_{h=1}^{n} \frac{1}{f(\mathbf{x}_h)} \sum_{t=1}^{k} h(z_t) \frac{\partial f(\mathbf{x}_h \mid z_j)}{\partial \alpha_{il}}. \end{aligned} \tag{6.9}$$

Now

$$\begin{aligned} \frac{\partial f(\mathbf{x}_h \mid z_t)}{\partial \alpha_{il}} &= f(\mathbf{x}_h \mid z_t) \frac{(-1)^{x}ih^{+1}}{\{\pi_i(z_t)\}^{x_{ih}} \{1 - \pi_i(z_t)\}^{1-x_{ih}}} \frac{\partial \pi_i(z_t)}{\partial \alpha_{il}} \\ &= f(\mathbf{x}_h \mid z_t) \left\{ \frac{x_{ih}}{\pi_i(z_t)} - \frac{(1 - x_{ih})}{1 - \pi_i(z_t)} \right\} \frac{\partial \pi_i(z_t)}{\partial \alpha_{il}}. \end{aligned} \tag{6.10}$$

The last step can easily be verified by substituting first $x_{ih} = 1$ and then

$x_{ih} = 0$. Substituting in (6.9) and interchanging the summations,

$$\frac{\partial L}{\partial \alpha_{il}} = \sum_{t=1}^{k} \frac{\partial \pi_i(z_t)}{\partial \alpha_{il}} h(z_t)$$

$$\times \left\{ \frac{\sum_{h=1}^{n} x_{ih} f(\mathbf{x}_h \mid z_t)/f(\mathbf{x}_h) - \sum_{h=1}^{n} \pi_i(z_t) f(\mathbf{x}_h \mid z_t)/f(\mathbf{x}_h)}{\pi_i(z_t)\{1 - \pi_i(z_t)\}} \right\}$$

$$= \sum_{t=1}^{k} \frac{\partial \pi_i(z_t)}{\partial \alpha_{il}} \frac{\{r_{it} - N_t \pi_i(z_t)\}}{\pi_i(z_t)\{1 - \pi_i(z_t)\}} \tag{6.11}$$

where

$$r_{it} = h(z_t) \sum_{h=1}^{n} x_{ih} f(\mathbf{x}_h \mid z_t)/f(\mathbf{x}_h) = \sum_{h=1}^{n} x_{ih} h(z_t \mid \mathbf{x}_h) \tag{6.12}$$

and

$$N_t = h(z_t) \sum_{h=1}^{n} f(\mathbf{x}_h \mid z_t)/f(\mathbf{x}_h) = \sum_{h=1}^{n} h(z_t \mid \mathbf{x}_h) \tag{6.13}$$

where $h(z_t \mid \mathbf{x}_h)$ is the posterior probability of z_t given $\mathbf{x}_h$.

It may be helpful to give an interpretation of r_{it} and N_t before describing how (6.11) can be made the basis of an E-M algorithm. The quantity $h(z_t \mid \mathbf{x}_h)$ is the probability that an individual with response vector $\mathbf{x}_h$ is located at z_t; $\sum_{t=1}^{k} h(z_t \mid \mathbf{x}_h)$ is thus the expected number of individuals at z_t. By a similar argument r_{it} is the expected number of those predicted to be at z_t who will respond positively.

The point of introducing r_{it} and N_t is to bring out the formal equivalence between (6.11) and the estimation equations which arise in dosage response experiments. In that case $\pi_i(z_t)$ would be the probability of an individual responding at dosage z_t, N_t would be the number treated at that level and r_{it} the number responding. There are well-known methods of probit and logit analysis which yield efficient estimators based on weighted linear regression for this problem. When the actual location of each individual on the z-scale is known, N_t is a known number and r_{it} an observed frequency. The estimation of the parameters is then simply a matter of solving the equations arising from (6.11) with $l = 1$ and 2.

In our case, N_t and r_{it} are functions of the unknown parameters, so we define an E-M algorithm as follows:

Step 1 Choose starting values for the α's.
Step 2 Compute the values of r_{it} and N_t from (6.12) and (6.13).

Step 3 Obtain improved estimates of the α's by solving (6.11) for $l = 1, 2, i = 1, 2, \ldots, p$, treating r_{it} and N_t as given numbers.

Step 4 Return to Step 2 and continue until convergence is attained.

Bock and Aitkin (1981) proposed this method to obtain estimates when z was, in fact, continuous. This involves choosing the number and location of the nodes $z_1, z_2, \ldots, z_k$ so that the various sums over t approximate the corresponding integrals that arise in the continuous time treatment. This was achieved using Gauss–Hermite quadrature routines available in the NAG(1986) FORTRAN software library. These authors reported that adequate solutions could be obtained with small numbers of nodes, for example $k = 3$, 5 or 7 and this would make it feasible to generalize the method to several latent variables. However, more recent and detailed investigations by Shea (1984) show that many more nodes (at least 20) may be necessary to get reasonable accuracy, and this places much greater demands on computing resources.

Nevertheless, as our development of the method shows, it can be regarded not simply as an approximation to a continuous variable problem, but as a very general method in its own right. It will work for any discrete prior distribution and for any response function. Bock and Aitkin (1981) used the normal function for $\pi_i(z_t)$ but the equations are much simpler if we use the logit. Thus, if

$$\pi_i(z) = 1/\{1 + \exp(-\alpha_{i0} - \alpha_{i1}z)\}$$

then

$$\frac{\partial \pi_i(z)}{\partial \alpha_{il}} = z^{l-1}\pi_i(z)\{1 - \pi_i(z)\} \quad (l = 1, 2). \tag{6.14}$$

The equations of (6.11) thus become

$$\sum_{t=1}^{k} z_t^{l-1} \sum_{h=1}^{n} \{x_{ih} - \pi_i(z_t)\} h(z_t \mid \mathbf{x}_h) = 0 \quad \begin{Bmatrix} l = 1, 2 \\ i = 1, 2, \ldots, p \end{Bmatrix} \tag{6.15}$$

which may be compared with (6.5).

In the first version of the method, individuals could have any value of z and at the E-step we predicted a value for each individual. The set of predictions $\{E(z \mid \mathbf{x}_h)\}$ could be thought of as constituting an empirical estimate of the distribution of z for the whole sample. In the second version the set of values of z which can occur is fixed and we have to predict how many individuals are located at each z. The set of values $\left\{\sum_{h=1}^{n} h(z_t \mid \mathbf{x}_h)\right\}$ thus also constitutes an estimate of the same distribution of z. In a rough sense the latter can be regarded as a grouped version of the former.

Note that in both versions it is only necessary to evaluate posterior probabilities and expectations for those score patterns that actually occur. Further, these quantities will be the same for all individuals with the same score pattern. Hence the number of evaluations to be made is equal to the number of distinct score patterns which will often be much less than 2^p.

Generalized Least Squares Methods

Before the introduction of the E-M algorithm for obtaining maximum likelihood estimators two other methods were devised for circumventing the heavy calculations required for full maximum likelihood. These were based on weighted least squares, and as they have the advantage of being applicable where there are several latent variables we now give a brief outline of them. They were proposed for the probit model but the idea could readily be extended to the other members of the family.

Both methods start with the presumption that most of the relevant information in the sample data is contained in the first- and second-order margins. An heuristic justification for this is obtained by considering the UV model. If the ξ's were known then we know that the sample dispersion matrix is sufficient for the model parameters, and these require only a knowledge of the bivariate distributions. One would expect the bivariate distributions of the indicator variables, $\mathbf{x}$, to contain almost all of the information about the underlying bivariate distributions. This is supported by an approximation to the likelihood function, given below, which depends only on the first- and second-order margins. Retrospectively, of course, one can see that the sampling errors of the least squares estimators are virtually the same as those obtained by maximum likelihood.

Christofferson's (1975) method was to choose parameter estimates which minimize the distance, in a least squares sense, between the observed and expected first- and second-order marginal proportions. Let $\hat{P}_i$ be the observed proportion who respond positively on variable i, and $\hat{P}_{ij}$ the proportion for variables i and j with P_i and P_{ij} being the corresponding population values. The parameters are then estimated by minimizing

$$SS = (\hat{\mathbf{P}} - \mathbf{P})'\mathbf{\Sigma}_P^{-1}(\hat{\mathbf{P}} - \mathbf{P}) \tag{6.16}$$

where $\mathbf{P}' = (P_1, P_2, \ldots, P_p, P_{12}, P_{13}, \ldots, P_{1p}, \ldots, P_{p,p-1})$ (of dimension $p + \frac{1}{2}p(p-1)$) and $\mathbf{\Sigma}_P$ is the dispersion matrix of P. Christofferson (1975) showed that a consistent estimator of $\mathbf{\Sigma}_P$ could be obtained and hence that efficient estimators could be found. ($\mathbf{\Sigma}_P$ involves third- and fourth-order marginal proportions.)

Although this method is much faster than full maximum likelihood Muthén (1978) showed that a further improvement was possible. At each stage of the iterative maximization of (6.16) one has to evaluate the integrals

$$P(\tau_i) = \int_{-\infty}^{\tau_i} \phi(u)\,\mathrm{d}u,\; P(\tau_i, \tau_j) = \int_{-\infty}^{\tau_i}\int_{-\infty}^{\tau_j} \phi(u_1, u_2, ; l_{ij})\,\mathrm{d}u_1\,\mathrm{d}u_2 \quad (6.17)$$

where ϕ is the standard univariate (or bivariate) normal density function. Muthén's (1978) method was to invert the equations (6.17) to give

$$\tau_i = \Phi^{-1}(P(\tau_i)),\; l_{ij} = f(P(\tau_i, \tau_j), P(\tau_i), P(\tau_i), P(\tau_j)), \quad \text{say}$$
$$(i, j = 1, 2, \ldots, p; i \neq j), \quad (6.18)$$

and then to fit the model by minimizing the weighted squared distance between the parameters of (6.18) and their sample estimates. To obtain the weights Muthén used a Taylor expansion expressing the estimator $\hat{t}$ of $\mathbf{t} = (\tau_1, \ldots, \tau_p, l_{12}, \ldots, l_{p-1,p})$ in the form

$$\hat{\mathbf{t}} = \mathbf{t} + \boldsymbol{\delta}$$

showing that the dispersion matrix of $\boldsymbol{\delta}$ could be consistently estimated. This method avoids the repeated integration involved in Christofferson's approach while retaining the same asymptotic efficiency. The sample estimators of $\{\hat{l}_{ij}\}$ are, of course, the tetrachoric correlations, so the method is analogous to the weighted least squares method for normal models. The difference lies in the use of tetrachoric correlations with consequent differences in the weighting.

6.2 Approximate methods based on correlations

A long-established method of factor analysing binary data has been to treat the tetrachoric correlations as if they were product-moment correlations and then to use a standard factor analysis program. This amounts to replacing Muthén's weight matrix with the unit matrix, so it might be described as an unweighted least squares method. Apart from its lack of efficiency, this method has the practical disadvantage, already noted, that the correlation matrix may not be positive semi-definite. This means that there will be at least one negative eigenvalue. This may not be practically important if any negative eigenvalues are small in absolute value, and this minor disadvantage has to be set against the very substantial computational advantages. We shall therefore pursue the question of how far approximate methods based on factor analysing pseudo-correlation coefficients may be useful.

Hitherto such methods have been discussed in the context of the probit model. Here we shall treat them in a larger framework. Their great advantage is that there is no serious restriction on the number of variables, manifest or latent, that can be handled.

For any underlying variable model of the form (5.12) the correlation coefficient between ξ_i and ξ_j is

$$\rho_{ij} = \sum_{h=1}^{q} \lambda_{ih}\lambda_{jh} \Big/ \Big(\sum_h \lambda_{ih}^2 + \psi_i\Big)^{\frac{1}{2}} \Big(\sum_h \lambda_{jh}^2 + \psi_j\Big)^{\frac{1}{2}}$$

$$= \sum_{h=1}^{q} \alpha_{ih}\alpha_{jh} \Big/ \Big(\sum \alpha_{ih}^2 + 1\Big)^{\frac{1}{2}} \Big(\sum \alpha_{jh} + 1\Big)^{\frac{1}{2}} \qquad (6.19)$$

by (5.18). Thus if we can obtain estimates of these correlation coefficients using the indicator variables $\{x_i\}$ then least squares or some other method can be used to provide estimators of $\{\alpha_{ih}\}$ $(i = 1, 2, \ldots, p; h = 1, 2, \ldots, q)$. The parameters $\{\alpha_{i0}\}$, $(i = 1, 2, \ldots, p)$ can then be estimated from the first-order marginal proportions. We have seen that if $\mathbf{z}$ and $\mathbf{e}$ are both normal then the tetrachoric coefficients provide the required estimators. Even if $\mathbf{z}$ and $\mathbf{e}$ are not normal the fact that $\boldsymbol{\xi}$ is a linear combination of independent random variables will make it approximately normal by the central limit theorem. However, if $q = 1$ and if $\mathbf{z}$ or $\mathbf{e}$ were markedly non-normal there might be better ways of estimating the correlation coefficients. There are, in fact, many other correlation coefficients which can be calculated from 2×2 contingency tables besides the tetrachoric coefficients, and most have the advantage of being much easier to calculate. It is therefore worth considering the problem from a slightly different angle following Chambers (1982).

To motivate what follows we consider the justification for using tetrachoric coefficients starting at the other end. That is we ask "What would the underlying bivariate distribution of ξ_i and ξ_j have to be for the tetrachoric coefficient to be reasonable?" Any coefficient has to be a function of $P(\tau_i, \tau_j)$, $P(\tau_i)$ and $P(\tau_j)$ (defined in (6.17)) since these are the only quantities which can be estimated. The tetrachoric coefficient is defined implicitly by (6.17) and these equations imply that, provided ϕ is bivariate normal, all possible values of the three quantities will yield the same ρ_{ij}; that is, the bivariate normal distribution has the property that whatever values of τ_i and τ_j are chosen the value of ρ_{ij} will be the same. We might therefore describe it as the "constant tetrachoric distribution". Now since τ_i and τ_j are arbitrary this is precisely the condition that we would want the distribution to have. Having reached this point we would go on to argue that since the bivariate normal distribution also results from a linear model like (5.12) the model provides the basis for interpreting a factor analysis of the tetrachoric correlations.

We now explore the consequences of starting with any other convenient correlation coefficient. Take, for example the so-called phi-coefficient, r_ϕ, which is the product moment coefficient obtained by treating the category codes 0 and 1 as variate values. It follows that the "constant r_ϕ" distribution must be given by

$$P(x, y) = P(x)P(y) + r_\phi[P(x)P(y)\{1 - P(x)\}\{1 - P(y)\}]^{\frac{1}{2}}. \quad (6.20)$$

For this to be a legitimate bivariate distribution function it must be monotonic increasing in x for fixed y. Since $P(x)$ is monotonic increasing in x, $P(x, y)$ must also be monotonic increasing in $P(x)$ for given y, and this is clearly not the case unless $r_\phi = 0$. Thus there is no continuous bivariate distribution with the required property and therefore no interpretation of a factor analysis of phi-coefficients which can be made in terms of a linear model.

Most of the other coefficients of correlation for 2×2 tables are functions of the cross-product (or, odds) ratio. If the array of frequencies is denoted by

$$\begin{matrix} a & b \\ c & d \end{matrix}$$

the cross-product ratio is defined as $\gamma = ad/bc$. Distributions for which γ is constant were introduced in Section 5.3 where they were called C-type distributions. For any member of this family the correlation coefficient will be a function of γ only. Suitable coefficients for factor analysis can then be devised by choosing appropriate forms for $P(x)$ and $P(y)$, the marginal distributions. Since we only observe $P(x)$ and $P(y)$ at a single point there is nothing the data can tell us about the form of the distributions and it is not clear how a choice may be made. Fortunately the choice does not seem to be critical. With rectangular margins

$$r_\gamma = \frac{\gamma + 1}{\gamma - 1} - \frac{2\gamma \ln \gamma}{(\gamma - 1)^2}. \quad (6.21)$$

Chambers (1982) shows that correlation coefficients derived from this family are well approximated by a coefficient of the form

$$r_\nu = \frac{\gamma^\nu - 1}{\gamma^\nu + 1}. \quad (6.22)$$

For rectangular margins $\nu = 0.67$ is a good approximation and for normal margins $\nu = 0.74$.

The mere existence of a suitable underlying bivariate distribution is not sufficient to justify a factor analysis based on r_v. It is also necessary that it should result from a model, preferably linear as in (5.12), for some distribution of $\mathbf{z}$ and $\mathbf{e}$. No such model appears to be known. However, we know that (ξ_i, ξ_j) is likely to be approximately normal, and Plackett (1965) and Mosteller (1968) have shown that a C-type distribution with normal margins is very close to this. (The same is true for the C-type logistic distribution, see Fachel (1986)). A good approximation should thus be obtained by factor analysing r_v coefficients with $v = 0.74$. Fachel (1986) has shown that the choice of v is not critical.

The loadings estimated by this means are not the α's themselves but, as (6.19) shows, simply related to them. If the parameters we estimate are denoted by β_{ih} then

$$\beta_{ih} = \alpha_{ih} \Big/ \left(1 + \sum_{h=1}^{q} \alpha_{ih}^2\right)^{\frac{1}{2}}. \tag{6.23}$$

Hence

$$\hat{\alpha}_{ih} = \hat{\beta}_{ih} \Big/ \left(1 - \sum_{h=1}^{q} \hat{\beta}_{ih}^2\right)^{\frac{1}{2}}, \quad (i = 1, 2, \ldots, p; h = 1, 2, \ldots, q). \tag{6.24}$$

Having estimated α_{ih}, $(i = 1, 2, \ldots, p; h = 1, 2, \ldots, q)$, it remains to estimate α_{i0}, $(i = 1, 2, \ldots, p)$. This is most easily done by equating the observed and expected marginal proportions. Thus the left-hand side of

$$Pr\{x_i = 1\} = \int_{-\infty}^{+\infty} \!\!\cdots\! \int \pi_i(\mathbf{z})\phi(\mathbf{z})\, d\mathbf{z} \tag{6.25}$$

can be estimated by the proportion responding positively on variable i, and we then choose α_{i0} in the right-hand side so that equality holds. This requires the numerical solution of the implicit equation, but a very good approximation for the logit/probit model is obtained as follows. If P_i is the observed proportion,

$$P_i = \int_{-\infty}^{+\infty} \!\!\cdots\! \int \frac{1}{\{1 + \exp(-\alpha_{i0} - \sum \alpha_{ij} z_j)\}} \phi(\mathbf{z})\, d\mathbf{z}. \tag{6.26}$$

The logistic distribution is close to the normal with variance $\pi^2/3$ so

$$1 + \exp\left(-\alpha_{i0} - \sum_j \alpha_{ij} z_j\right) \doteqdot \Phi\left\{\frac{\alpha_{i0} + \sum_j \alpha_{ij} z_j}{\pi/\sqrt{3}}\right\}.$$

Substitution into (6.26) and a change of variable gives

$$P_i = \Phi\left\{\alpha_{i0} \Big/ \left(1 + \frac{3}{\pi^2} \sum_j \alpha_{ij}^2\right)^{\frac{1}{2}}\right\}. \tag{6.27}$$

Hence we take as our estimator

$$\hat{\alpha}_{i0} = \left(1 + \frac{3}{\pi^2} \sum_j \hat{\alpha}_{ij}^2\right)^{\frac{1}{2}} \Phi^{-1}(\hat{P}_i) \tag{6.28}$$

or

$$\hat{\pi}_i = 1/(1 + e^{-\alpha_{i0}}), \quad (i = 1, 2, \ldots, p).$$

6.3 Approximate methods based directly on cross-product ratios

A second way of obtaining approximate estimators stems from a result of Bartholomew (1980) for the model when G^{-1} is the logit function. This starts with the cross-product ratios in the expectation that they contain most of the information about the associations between the variables. If the α's are small the theoretical cross-product ratios can be expanded in a Taylor series the first few terms of which may serve as an approximation.

The expected proportions of individuals falling in the cells of the 2×2 table for variables i and j will be

$$\begin{array}{cc} Ex_i x_j & Ex_i(1-x_j) \\ E(1-x_i)x_j & E(1-x_i)(1-x_j). \end{array}$$

The cross-product ratio for this table is defined to be

$$\gamma_{ij} = Ex_i x_j E(1-x_i)(1-x_j)/Ex_i(1-x_j)E(1-x_i)x_j. \tag{6.29}$$

Using the fact that $Ex_i = E\pi_i(\mathbf{y})$, $Ex_i x_j = E\pi_i(\mathbf{y})\pi_j(\mathbf{y})$, etc., and expanding as a power series in the α's, we find that

$$\gamma_{ij} = 1 + \sigma^2 \sum_{h=1}^{q} \alpha_{ih}\alpha_{jh} + \text{terms of 4th degree}, \; (i \neq j = 1, 2, \ldots, p) \tag{6.30}$$

where $\sigma^2 = E\{H^{-1}(y_j)\}^2 = E(z_j^2)$ for all j. We have hitherto parametrized the model so that $\sigma^2 = 1$, which gives

$$\gamma_{ij} - 1 \doteqdot \sum_{h=1}^{q} \alpha_{ih}\alpha_{jh}, \quad (i \neq j = 1, 2, \ldots, p). \tag{6.31}$$

The right-hand expression is identical in form with that for the correlations in the linear factor model. This suggests that the model might be fitted by treating the $(\gamma_{ij} - 1)$'s as correlations. It turns out that (6.31) is a good approximation when $q = 1$ but rather poor for $q > 1$, so it should only be used for fitting one-factor models. The accuracy may be judged from the figures in Table 6.1 which should all be close to 1. The worst cases occur when α_{i1} and α_{j1} are very unequal and the reason for this will appear later. In practice the α's are often

Table 6.1 Values of $(\gamma_{ij}-1)/\alpha_{i1}\alpha_{j1}$ for the logit model. The entries are unchanged if (π_i, π_j) is replaced by $(1-\pi_i, 1-\pi_j)$ and if α_{i1} and α_{j1} are interchanged.

$(\alpha_{i1}\alpha_{j1})$ \ (π_i, π_j)	$(\frac{1}{2}, \frac{1}{2})$	$(\frac{1}{4}, \frac{3}{4})$	$(\frac{1}{4}, \frac{1}{4})$	$(\frac{1}{10}, \frac{1}{10})$	$(\frac{1}{20}, \frac{1}{20})$
(2, 2)	0.942	1.192	0.984	1.119	1.280
(2, 1)	0.801	0.944	0.850	1.008	1.196
$(2, \frac{1}{2})$	0.614	0.668	0.644	0.731	0.820
(1, 1)	0.912	1.063	0.988	1.245	1.576
$(1, \frac{1}{2})$	0.846	0.934	0.912	1.125	1.372
$(\frac{1}{2}, \frac{1}{2})$	0.935	1.011	1.015	1.263	1.535
$(\frac{1}{2}, \frac{1}{4})$	0.917	0.965	0.977	1.139	1.288
$(\frac{1}{4}, \frac{1}{4})$	0.971	1.001	1.016	1.119	1.188
$(\frac{1}{10}, \frac{1}{10})$	0.994	1.000	1.004	1.018	1.003

close together and well within the range of values spanned by the table. Even if the approximate estimates to which they lead are not adequate, they provide good starting values for more efficient iterative methods.

We now give three quick methods of fitting the one-factor logit model based on (6.31). All could also be used for fitting the standard factor model.

Method I

This depends on the fact that if we equate the observed and expected values of

$$\sum_{\substack{j=1\\j\neq i}}^{p} (\gamma_{ij}-1), \quad (i=1, 2, \ldots, p)$$

we have just enough equations to determine the α's uniquely. The problem, then, is to solve the system of equations

$$\alpha_{i1}\sum_{\substack{j=1\\j\neq i}}^{p} \alpha_{j1} = \sum_{\substack{j=1\\j\neq i}} (\hat{\gamma}_{ij}-1) = \hat{T}_i, \text{ say}, \quad (i=1, 2, \ldots, p) \qquad (6.32)$$

where $\hat{\gamma}_{ij}$ is the sample estimator of γ_{ij}. One method of solving the equations was given in Bartholomew (1980). Another, well suited to a microcomputer, is based on the iterative formula

$$\alpha_{i1}^{(r+1)} = \{\hat{T}_i + (\alpha_{i1}^{(r)})^2\} \Big/ \left\{\left|\sum_{i=1}^{p} \hat{T}_i\right| + \sum_{i=1}^{p} (\alpha_{i1}^{(r)})^2\right\}^{\frac{1}{2}}. \qquad (6.33)$$

Before using the method the categories should be ordered so that as many as possible of the γ_{ij}'s are greater than one. To start, the α's may all be taken equal to 0.5, or 1. For a poorly fitting model there may be no real solution of (6.32). The α_{i0}'s (or π_i's) may be estimated by the formulae of (6.28).

Method II

This is essentially the "minres" method of Harman and Jones (1966) and is based on minimizing

$$SS = \sum_{\substack{i=1 \\ i \neq j}}^{p} \sum_{j=1}^{p} (\hat{\gamma}_{ij} - 1 - \alpha_{i1}\alpha_{j1})^2. \tag{6.34}$$

The equations obtained by setting the partial derivatives equal to zero can be solved iteratively using

$$\alpha_{i1}^{(r+1)} = \left\{ \sum_{j=1}^{p} (\hat{\gamma}_{ij} - 1)\alpha_{i1}^{(r)} + (\alpha_{i1}^{(r)})^3 \right\} \Big/ \sum_{i=1}^{p} (\alpha_{i1}^{(r)})^2. \tag{6.35}$$

Starting values of $\alpha_{i1} = 1$ for all i are usually satisfactory. This method of solution is superior to those given by Comrey (1962) and Comrey and Ahumada (1964).

Method III

A method involving no iteration was suggested by Shea (1984) to obtain starting values for the E-M algorithm. If $\gamma_{ij} - 1 \doteqdot \alpha_{i1}\alpha_{j1}$ then

$$(\gamma_{ij} - 1)(\gamma_{jk} - 1)/(\gamma_{ik} - 1) \doteqdot \alpha_{j1}^2.$$

For each pair (i, k) we can then obtain an estimate of α_{j1}. A simple average over all pairs (excluding any for which $\gamma_{ik} = 1$) then gives

$$\hat{\alpha}_{j1}^2 = \sum_{\substack{i=1 \\ i,k \neq j}}^{p-1} \sum_{k=i+1}^{p} \frac{2(\hat{\gamma}_{ij} - 1)(\hat{\gamma}_{jk} - 1)}{(\hat{\gamma}_{ik} - 1)(p-1)(p-2)}. \tag{6.36}$$

There is an ambiguity of sign with all these estimates since the signs of all α's may be changed without affecting the fit or the interpretation.

An Alternative Justification of the Taylor Series Approximation

It seems surprising that an approximation based on a Taylor expansion which requires the α's to be small should be as good as Table 6.1 shows it to be for quite large α's. Some insight into the reasons for this can be gained by relating it to the correlation approach as follows.

According to the main result of Section 5.4 the logit model is equivalent to a UV model in which the error term, e_i, has a logistic distribution with variance $\psi_i\pi^2/3$. It follows that

$$\begin{aligned}\rho(\xi_i, \xi_k) &= \frac{\lambda_{i1}\lambda_{k1}}{\{(\lambda_{i1}^2 + \psi_i\pi^2/3)(\lambda_{k1}^2 + \psi_{k1}\pi^2/3)\}} = \frac{\alpha_{i1}\alpha_{k1}}{\{(\alpha_{i1}^2 + \pi^2/3)(\alpha_{k1}^2 + \pi^2/3)\}} \\ &= \alpha_{i1}\alpha_{k1}/\{\alpha_{i1}^2\alpha_{k1}^2 + (\alpha_{i1} - \alpha_{k1})^2\pi^2/3 + 2\alpha_{i1}\alpha_{k1}\pi^2/3 + \pi^4/9\} \\ &= \alpha_{i1}\alpha_{k1}/\{(\alpha_{i1}\alpha_{k1} + \pi^2/3)^2 + (\alpha_{i1} - \alpha_{k1})^2\pi^2/3\}^{\frac{1}{2}} \\ &\geqslant \alpha_{i1}\alpha_{k1}/(\alpha_{i1}\alpha_{k1} + \pi^2/3). \qquad (6.37)\end{aligned}$$

Now if $\alpha_{i1} - \alpha_{k1}$ is small, the correlation will be close to its lower bound as given by (6.37), and since $\alpha_{i1}\alpha_{k1} \doteqdot \gamma_{ik} - 1$,

$$\rho(\xi_i, \xi_k) \doteqdot \alpha_{i1}\alpha_{k1}/(\alpha_{i1}\alpha_{k1} + \pi^2/3) \doteqdot (\gamma_{ik} - 1)/(\gamma_{ik} - 1 + \pi^2/3). \qquad (6.38)$$

For values of γ_{ik} which one is likely to meet in practice, this expression is close to other functions of the cross-product ratios that we have already seen are good approximations to the correlation coefficient of the underlying normal or C-type distribution. This is illustrated numerically in Table 6.2.

Table 6.2 Comparison of estimates of correlation based on cross-product ratios

γ	1	1.25	1.5	2	3	4	5	10	100
$(\gamma - 1)/(\gamma - 1 + \pi^2/3)$ (6.38)	0	.071	.132	.233	.378	.477	.549	.732	.968
$(\gamma^{.74} - 1)/(\gamma^{.74} + 1)$ (6.22)	0	.082	.149	.251	.385	.472	.534	.692	.936
$(\gamma + 1)/(\gamma - 1) - \dfrac{2\gamma \ln \gamma}{(\gamma - 1)^2}$ (6.21)	0	.074	.134	.227	.352	.434	.474	.654	.926

Since sample values of γ are commonly in the range 1 to 5, the correlation methods and those based on the Taylor series may be expected to give very similar results. One slight advantage of the method of this section is that it gives the α's directly without the need of the transformation of (6.24). The analysis also shows why the approximation breaks down if $q > 1$; it is not then possible to express the correlation as a function of the cross-product ratios as in (6.37).

6.4 Approximate maximum likelihood estimators

In certain circumstances it is possible to approximate the likelihood function in a manner which allows us to obtain the maximum likelihood estimators explicitly. As the result for binary data is a special case of a more general result for polytomous data to be given in Chaper 8 we merely quote the result here. The method starts with the matrix of phi-coefficients between all pairs of the p variables. We then find the eigenvectors associated with all eigenvalues greater than one. Let $\mathbf{g}$ be the normalized vector associated with the largest eigenvalue; then there is a vector $\boldsymbol{\alpha}' = (\alpha_{11}, \alpha_{21}, \ldots, \alpha_{p1})$ given by

$$\boldsymbol{\alpha} \propto \hat{\mathbf{Q}}^{-\frac{1}{2}}\mathbf{g} \tag{6.39}$$

where $\hat{\mathbf{Q}}$ is a diagonal matrix with ith diagonal element equal to $\hat{\pi}_i(1 - \hat{\pi}_i)$ which maximizes the approximate likelihood. The constant of proportionality is determined in a manner to be explained later. The vector associated with the second largest eigenvalue provides a set of estimators for a second factor, and so on for all eigenvalues exceeding one. There is an obvious similarity between this method and a principal components analysis of the phi-matrix. The latter would extract the eigenvectors of the same matrix. However, the methods give different loadings because of the matrix $\hat{\mathbf{Q}}$ and the constant of proportionality in (6.39). Like all of the appro mate methods this is concerned only with the α_{ij}'s for $j \geqslant 1$; the α_{i0}'s, or π_i's can easily be estimated using (6.26) or (6.28).

6.5 Breakdown of estimation procedures

As with the normal model, it sometimes happens that any of the estimation procedures will fail to give sensible answers. In the normal case we met this phenomenon when estimates of ψ_i turned out to be zero—known as Heywood cases. The analogous situation with binary models occurs when an iterative procedure fails to converge. When this happens it is usually because one of the estimators is becoming larger and larger. Recall that, in the UV model representation, $\alpha_{ij} = -\lambda_{ij}/\psi_i^{\frac{1}{2}}$, so a diverging alpha is equivalent to a ψ_i approaching zero. Experience with binary estimation procedures confirms that the circumstances conducive to this type of behaviour are similar to those which lead to Heywood cases. That is,

(a) if the sample size is small, a few hundred or less;
(b) if the number of variables is small;
(c) if the α's are very unequal.

In practice the iteration will cease when the convergence criterion of the algorithm or some other stopping rule is satisfied. The actual value of the estimate returned in such cases will thus be a function of the algorithm. For purposes of interpretation the precise value of the estimate is not crucial—what matters is that it is "large". One very large value of α_{i1}, for example, will dominate the component. We shall meet an example of this kind in Chapter 9 (see Table 9.7) where possible interpretations will be discussed. As one would expect, "large" estimates will also have large standard errors.

It is worth adding in connection with (a) above that there is some relevant theoretical work by Albert and Anderson (1984) which throws some further light on the problem. They were concerned with ordinal regression in which the regressor variables are not latent as in our case, but their results would apply to the M-step of the E-M algorithm and hence to the full solution. These authors identify the conditions under which divergence can occur and they show that the probability of this happening tends to zero as n tends to infinity.

6.6 Sampling properties

Sampling Variation of the Maximum Likelihood Estimators

The determination of exact standard errors for the parameter estimates does not appear to be possible. The usual way to deal with this situation for maximum likelihood and generalized least squares estimation is to compute the asymptotic variance–covariance matrix using the information matrix. Even this proves to be impracticable for large p and so we shall give a more tractable alternative (the same point was noted for latent class models in Section 2.2). In order to give some idea of the accuracy of the asymptotic formulae and to obtain some information about the comparative performance of the approximate methods of estimation we shall also give some estimated means and standard errors derived by simulation.

Asymptotically the sampling variances and covariances of the maximum likelihood estimates of the α's are given by the elements of the inverse of the information matrix evaluated at the solution point. Thus if we have a set of parameters $\boldsymbol{\beta}$ then

$$\{D(\hat{\boldsymbol{\beta}})\}^{-1} = E\left[-\frac{\partial^2 L}{\partial\beta_i\,\partial\beta_j}\right]_{\boldsymbol{\beta}=\hat{\boldsymbol{\beta}}}$$

in which

$$\frac{\partial^2 L}{\partial\beta_i\,\partial\beta_j} = \sum_{h=1}^{n}\left\{\frac{1}{f}\frac{\partial^2 f}{\partial\beta_i\,\partial\beta_j} - \frac{1}{f^2}\frac{\partial f}{\partial\beta_i}\frac{\partial f}{\partial\beta_j}\right\} \tag{6.40}$$

where $f \equiv f(\mathbf{x}_h)$. On taking the expectation the first term vanishes, leaving

$$\{D(\hat{\boldsymbol{\beta}})\}^{-1} = n\left\{E\frac{1}{f^2}\frac{\partial f}{\partial\beta_i}\frac{\partial f}{\partial\beta_j}\right\}_{\boldsymbol{\beta}=\hat{\boldsymbol{\beta}}}. \tag{6.41}$$

In our case $\mathbf{x}$ is a score pattern taking 2^p different values and the expectation in (6.41) is thus

$$\sum_{\text{all } \mathbf{x}} \frac{1}{f(\mathbf{x})} \frac{\partial f(\mathbf{x})}{\partial \beta_i} \frac{\partial f(\mathbf{x})}{\partial \beta_j}. \tag{6.42}$$

If p is small it is feasible to evaluate this sum for all i and j and then to invert the resulting matrix. However, if p is larger the computer resources required to compute (6.42) become prohibitive. Moreover many of the probabilities $f(\mathbf{x}_h)$ will become so small that the computation of $1/f(\mathbf{x}_h)$ will cause overflow on most computing machines. In these circumstances an approximation can be obtained by replacing the expectation of the information matrix by its observed value. This requires the computation of (6.40) and inversion of the resulting matrix. Since the first term has expectation zero a further approximation may be obtained from

$$D^*(\hat{\boldsymbol{\beta}}) = \left\{ \sum_{h=1}^{n} \frac{1}{f^2(x_h)} \frac{\partial f(\mathbf{x}_h)}{\partial \beta_i} \frac{\partial f(\mathbf{x}_h)}{\partial \beta_j} \right\}^{-1}. \tag{6.43}$$

The dimension of this matrix is $p + qp$ (p α_{i0}'s and p α_{ij}'s for the q values of j). The number of distinct terms in the sum of (6.43) will usually be less than n since more than one individual may have the same score pattern.

The approximation given by (6.43) appears to be good for standard errors but less reliable for covariances. Some comparative figures are given in Table 6.3 for standard errors and these are typical of the differences found in other cases. It therefore seems acceptable to use the simpler version of (6.43) which, in any case, tends to err on the safe side.

Table 6.3 Comparison of asymptotic and simulated means and standard errors for the logit model using LSAT VI data.

Parameter	Estimate	Standard error by (6.41)	Standard error by (6.43)	1000 replications Mean	1000 replications Standard deviation
α_{11}	.83	.25	.26	.85	.29
α_{21}	.72	.17	.19	.75	.22
α_{31}	.89	.20	.23	.93	.31
α_{41}	.69	.17	.19	.69	.20
α_{51}	.66	.19	.21	.66	.22
π_1	.94	.01	.01	.94	.01
π_2	.73	.02	.02	.73	.02
π_3	.56	.02	.02	.56	.02
π_4	.78	.01	.02	.79	.02
π_5	.89	.01	.01	.89	.01

Table 6.4 Comparison of asymptotic and simulated means and standard errors for the logit model using data from McHugh (1956)

Parameter	Estimate	Standard error by (6.43)	528 replications Mean	528 replications Standard deviation
α_{11}	1.17	.37	1.24	.34
α_{21}	1.06	.35	1.11	.31
α_{31}	3.78	2.14	5.57	5.44
α_{41}	2.15	.63	2.88	2.52
π_1	.44	.05	.44	.07
π_2	.50	.05	.50	.06
π_3	.32	.12	.32	.19
π_4	.44	.07	.44	.14

The next question concerns the adequacy of asymptotic sampling theory for finite samples. Some empirical evidence on this point is given in Tables 6.3 and 6.4. The left-hand part of each table gives the maximum likelihood estimates and the asymptotic standard errors for two typical data sets. The right-hand part gives results from simulation. Ideally we would want to obtain samples for the latter from a population having the same parameter values as the real data, but these, of course, are unknown. Instead we have sampled from a population with parameter values equal to the estimated value for the real data.

Table 6.3 contains results for data from the Law School Admission Test (LSAT VI) which will be discussed in more detail in Chapter 9. The sample size is 1000 and there are 5 variables. It is clear that the π's can be estimated without bias and with high precision. For the α's the bias, if any, appears to be small, but the standard errors are quite large. The asymptotic theory underestimates the true value, and the reason for this is apparent on inspection of the sampling distributions. These distributions are positively skewed and it is the occasional occurrence of an unusually large α which accounts for the larger true standard error. With increasing sample sizes such occurrences will be increasingly rare and the two will become closer.

This example is typical of what happens in large samples with similar α's of moderate size, such as occur in educational testing. The second example, in Table 6.4, was chosen because it combines a small sample size with fairly large and unequal α's.

The data relate to performance on engineering tests given in unpublished work by Schumacher, Maxson and Martinek and re-analysed by McHugh (1956). Again the π's are estimated without bias, though here we notice that the simulated standard errors are larger than the asymptotic values. This is because the estimates of α_{i1} and π_i are correlated and the differences here are a reflection of the bias in the estimates of α_{31} and α_{34}. The position for the other two α's is similar to the previous example, but there are large discrepancies in both the mean and standard error for the two large α's. This shows in greatly exaggerated form the effect of the skewness in the sampling distribution noted above. If one truncates the simulated sampling distribution at some arbitrary value, like 7, the actual and the asymptotic values would be much closer. In other words the asymptotic theory gives a good indication of the scatter of the estimates if we exclude the occasional extreme value to which the estimation method is prone. The important point to note here is that the asymptotic means and standard errors are likely to underestimate seriously the true values for α's much in excess of 1, especially in small samples.

The final question relates to the relative performance of the various approximate methods of estimation as compared with maximum likelihood. Since the former are very much easier to use than the latter it is important to know what is lost by using the simple alternatives. No general results are available and the following results are intended to be no more than indicative of directions for further research.

In Chapter 9 we shall give the results of applying different estimation methods to the same data (see Tables 9.3 and 9.6). The results are broadly similar with the main differences occurring when the α's are large. We notice, particularly, the tendency for maximum likelihood to produce larger estimates than the other methods. Since we have seen in Table 6.4 that this method has a tendency to overestimate we might surmise that the approximate methods will be less biased.

Fachel (1986) has carried out an extensive comparison of methods of estimation which confirm these conclusions, but she also draws attention to the fact that bias as such may not be important. What matters for interpretation is the *profile* of the α's. A systematic tendency to overestimate by the same factor, for example, would not affect the ranking of indviduals given by the components.

Our final table, Table 6.5, provides a comparison of the maximum likelihood method with the correlation approach described in Section 6.2 using simulation. The correlation coefficient used is $r_{0.74}$ as given by (6.22) and fitting is by the "minres" method. Each case is based on 100 replications for the case $n = 400$ and $p = 4$. Since the program used for the correlation method was designed for the logit/logit model the results are presented using the scaling for that version. To make them comparable with the logit/probit scaling all figures in the table must be multiplied by $\pi/\sqrt{3}$. Results are given only for the α's since all

Table 6.5 Means and standard deviations of sampling distributions of estimates of the α's for the logit/logit model

	Maximum likelihood		Minres on correlations	
α	Mean	Standard deviation	Mean	Standard deviation
.5	.53	.18	.50	.13
.5	.54	.18	.52	.15
.5	.52	.17	.55	.16
.5	.54	.20	.54	.17
1	1.00	.19	.98	.16
1	1.02	.22	.96	.16
1	1.04	.23	.97	.15
1	1.05	.18	.95	.16
0	.01	.08	.00	.07
1	1.03	.17	.97	.14
2	2.21	.62	1.78	.33
3	3.78*	3.10*	2.76	1.68
.5	.52	.14	.53	.10
.5	.53	.11	.54	.11
.5	.52	.11	.54	.11
3.0	6.80†	7.43†	2.26‡	1.35‡

* 4 cases failed to converge, results shown are based on 96 cases.
† 27 cases failed to converge
‡ 17 cases failed to converge.

methods estimate the π's precisely; the results are typical of those in a larger set from which these have been selected. We note again the upward bias of the maximum likelihood method especially where the α's are large. The most surprising feature is the greater precision of the minres method when compared with the asymptotically optimal maximum likelihood. If performance is judged by mean square error the conclusion is the same, although the differences are slightly smaller. The reason again appears to lie in the greater propensity of the maximum likelihood method to show divergent behaviour. It is a few very extreme cases which account for both the bias and the larger standard errors. If these are excluded the balance is redressed.

It must be emphasized again that this is only a limited exploratory investigation, but if the conclusions generalize there will be a strong case for using the approximate methods and developing appropriate sampling theory for them.

Goodness of Fit

If n is large compared with 2^p a chi-squared or log-likelihood goodness of fit test can be carried out in the usual way on the observed and expected frequencies of the score patterns. The distribution of frequencies over the categories is often such that there are many small expected frequencies so that grouping becomes necessary. This can sometimes lead to problems. The number of degrees of freedom in the ungrouped case is $2^p - p(q+1) - 1$. If the number of categories is reduced by grouping, there may easily be no degrees of freedom left to judge goodness of fit. If 2^p is of the same order as n, or greater, it will be impossible to base a test on a comparison of frequencies because forming categories with large enough frequencies will eliminate the degrees of freedom altogether. In these circumstances there is as yet no way of carrying out a formal test. It is nevertheless desirable to inspect the fit of the model. If 2^p is not too large this can be done comparing the observed and expected frequencies of the score patterns. However, this must be done with care especially if p is very large. For in that case all of the expected frequencies will be very small (of order 2^{-p}) whereas the observed frequencies will of necessity be integers and almost always 0 or 1. An additional check can be made by comparing the observed and fitted values of the one- and two-way marginal frequencies.

There are various other checks which can be made on the data either before or after fitting a model. For example it was shown in Bartholomew (1980) that if a one-factor model applies then it must be possible to label the categories so that all of the cross-product ratios exceed one. This is a "population" result, and sampling variation might cause violations of the inequalities, but as a general guide a one-factor model is unlikely to be adequate unless all of the cross-product ratios can be made greater than one. A systematic approach to the question of what patterns in the data are consistent with a one-factor model has been developed by Holland (1981) and extended by Rosenbaum (1984). They give theorems establishing inequalities of various kinds which must be satisfied by the one-factor model. Rosenbaum also illustrates the use of some of these results.

CHAPTER 7

Models for Polytomous Data

7.1 The response function approach

The general approach to choosing factor models set out in Chapter 4 yields a family of models for polytomous data which generalize those for binary data in Chapter 5. We first need to extend the notation used there. Let c_i denote the number of categories of variable i which are labelled $0, 1, \ldots, c_i - 1$ $(i = 1, 2, \ldots, p)$ and indexed by s. The indicator variable x_i is now replaced by a vector-valued indicator function defined by

$$\begin{aligned} x_i(s) &= 1 \quad \text{if the response falls in category } s \\ &= 0 \quad \text{otherwise.} \end{aligned}$$

The c_i-vector with these elements is denoted by $\mathbf{x}_i$ and, obviously, $\sum_s x_i(s) = 1$. The full response pattern for an individual is denoted by $\mathbf{x}' = (\mathbf{x}_1', \mathbf{x}_2', \ldots, \mathbf{x}_p')$ of dimension $\sum_i c_i$. Note that if we specialize to the binary case by putting $c_i = 2$ for all i, this notation does not coincide exactly with that in Chapter 5. The two-element vector $\mathbf{x}_i$ was there represented by the single element which here is denoted by $x_i(1)$. Since $x_i(0) + x_i(1) = 1$, it is sufficient to record only $x_i(1)$ and to delete alternate elements in $\mathbf{x}$; in the general case we could delete $x_i(0)$ for each i. However, the symmetry which results from retaining the redundancy implied by this notation outweighs any advantage of the more parsimonious description.

The single response function $\pi_i(\mathbf{y})$ is now replaced by a set of functions defined by

$$Pr\{x_i(s) = 1\} = \pi_{is}(\mathbf{y}), \quad (s = 0, 1, \ldots, c_i - 1; i = 1, 2, \ldots, p). \tag{7.1}$$

The requirement that $\sum_s \pi_{is}(\mathbf{y}) = 1$ again means that there is some redundancy in the notation. We may now suppose the conditional probability function of $\mathbf{x}_i$ given $\mathbf{y}$ to be multinomial so that

$$g_i(\mathbf{x}_i \mid \mathbf{y}) = \prod_{s=0}^{c_i-1} \{\pi_{is}(\mathbf{y})\}^{x_i(s)} \quad (i = 1, 2, \ldots, p). \tag{7.2}$$

The posterior density function of $\mathbf{y}$ is then given by

$$h(\mathbf{y} \mid \mathbf{x}) = h(\mathbf{y}) \prod_{i=1}^{p} \prod_{s=0}^{c_i-1} \{\pi_{is}(\mathbf{y})\}^{x_i(s)}/f(\mathbf{x})$$

$$= h(\mathbf{y}) \exp \sum_{i=1}^{p} \sum_{s=0}^{c_i-1} x_i(s) \log \pi_{is}(\mathbf{y})/f(\mathbf{x}). \qquad (7.3)$$

Reference to the general argument of Section 4.2 then indicates a model of the form

$$\log \pi_{is}(\mathbf{y}) = \alpha_{i0}(s) + \sum_{j=1}^{q} \alpha_{ij}(s) H^{-1}(y_j) + \text{a term not depending on } s \qquad (7.4)$$

where, as before, the y's are uniformly distributed on $(0, 1)$. The last term on the right-hand side is needed to ensure that the $\pi_{is}(y)$'s sum to one for each i. The fact that it does not depend on s ensures that when (7.4) is substituted into (7.3) $h(\mathbf{y} \mid \mathbf{x})$ does not depend on $x_i(s)$ other than through the components which are given by

$$X_j = \sum_{i=1}^{p} \sum_{s=0}^{c_i-1} \alpha_{ij}(s) x_i(s), \quad (j = 1, 2, \ldots, q). \qquad (7.5)$$

For many purposes it will be convenient to express the response function in terms of $z_j = H^{-1}(y_j)$, $(j = 1, 2, \ldots, q)$.

The requirement that $\sum_s \pi_{is}(\mathbf{y}) = 1$ together with (7.4) implies that

$$\pi_{is}(\mathbf{y}) = \exp\left\{\alpha_{i0}(s) + \sum_{j=1}^{q} \alpha_{ij}(s) H^{-1}(y_j)\right\} \bigg/ \sum_{r=0}^{c_i-1} \exp\left\{\alpha_{i0}(r) + \sum_{j=1}^{q} \alpha_{ij}(r) H^{-1}(y_j)\right\} \quad (i = 1, 2, \ldots, p; s = 0, 1, \ldots, c_i - 1). \qquad (7.6)$$

Fig. 7.1 illustrates the forms of the response functions for the set of categories on a given manifest variable for a single latent variable measured on the z-scale. As we move from left to right along the latent dimension the probability of a positive response is decreasing in some categories and increasing in others. The model thus determines an ordering of the categories such that the higher a subject is on the latent scale the greater is the tendency for the response to fall in the higher categories. This behaviour is plausible in many contexts. It would be appropriate, for example, if the categorization were arrived at by grouping individuals according to an imperfect perception of

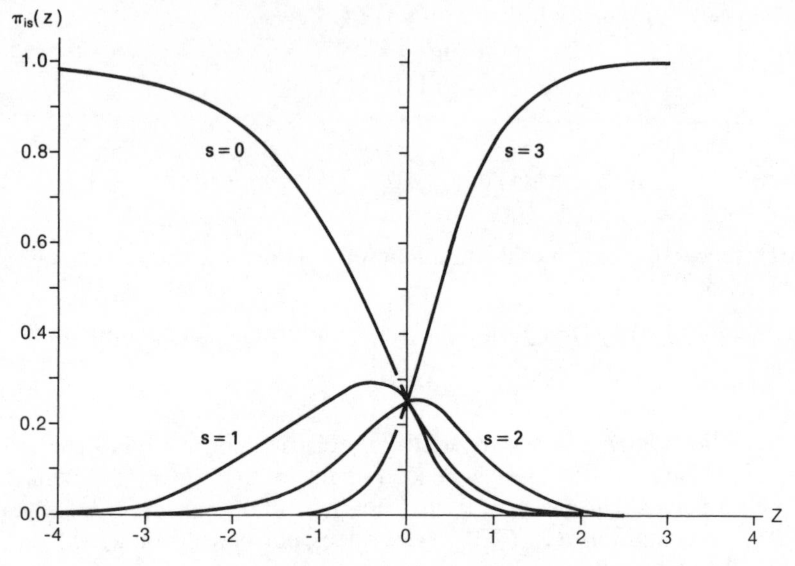

Fig. 7.1 $\pi_i(s) = \frac{1}{2}$ **for all** s **and** $\alpha_i(0) = 0$, $\alpha_i(1) = 1$, $\alpha_i(2) = 2$, $\alpha_i(3) = 4$

their place on the scale. In educational testing it would be plausible if the possible answers were of varying degrees of "rightness". However, as Thissen and Steinberg (1984) point out, this would not be the case where one response is correct and the others equally wrong. It would not then be reasonable to suppose that the weakest candidates would be distinguished by the selection of one particular wrong answer. The point made in connection with binary models bears reiteration: *one should not assume that the possession of attractive mathematical properties will guarantee the applicability of a model in any particular circumstances.* Thissen and Steinberg (1984) discuss alternative methods.

As expressed in (7.6), $\pi_{is}(\mathbf{y})$ is over-parametrized. This is easily demonstrated by replacing $\alpha_{ij}(s)$ by $\alpha_{ij}(s) + K_{ij}$ for any (i, j). This merely introduces a common factor into the numerator and denominator and so leaves the response probability unchanged. Without loss of generality we may therefore fix the location of the set $\{\alpha_{ij}(s)\}$, as we please. In the binary case we implicitly selected $\alpha_{ij}(0) = 0$. We shall do the same here, but later an alternative convention will offer advantages. The order in which the categories are labelled is arbitrary, but the one labelled 0 will be called the *reference* category for that variable. For most purposes it will prove to be simpler to express the model in terms of response probabilities referred to the reference category; thus

$$\log\{\pi_{is}(\mathbf{y})/(\pi_{i0}(\mathbf{y})\} = \alpha_{i0}(s) + \sum_{j=1}^{q} \alpha_{ij}(s)H^{-1}(y_j) \tag{7.7}$$

$$= \alpha_{i0}(s) + \sum_{j=1}^{q} \alpha_{ij}(s)z_j \tag{7.8}$$

$$(s = 1, 2, \ldots, c_i - 1; i = 1, 2, \ldots, p).$$

We shall extend the customary usage by referring to the function on the left-hand side as a logit and hence to (7.6), (7.7) or (7.8) as the logit model. It no longer makes sense to talk of a probit model since there is no simple comparable way of introducing a normal function on the left-hand side. The logit model was first proposed by Bock (1972) though with a rather different rationale.

An alternative parametrization which is useful later and which facilitates the interpretation of the $\alpha_{i0}(s)$'s is obtained as follows. If we put $z_j = H^{-1}(y_j) = 0$ for all j, we obtain the response probability for the "median" individual. Let this be denoted by

$$\pi_{is} = \exp \alpha_{i0}(s) \Big/ \sum_{r=0}^{c_i-1} \exp \alpha_{i0}(r). \tag{7.9}$$

Since the origin of the α's is arbitrary we may again set $\alpha_{i0}(0) = 0$ and then

$$\pi_{is} = \exp \alpha_{i0}(s) \Big/ \left\{1 + \exp \sum_{r=0}^{c_i-1} \alpha_{i0}(r)\right\} \quad (s = 1, 2, \ldots, c_i - 1). \tag{7.10}$$

Substituting into (7.6) then gives

$$\pi_{is}(\mathbf{z}) = \pi_{is} \exp \sum_{j=1}^{q} \alpha_{ij}(s)z_j \Big/ \sum_{r=0}^{c_i-1} \pi_{ir} \exp \sum_{j=1}^{q} \alpha_{ij}(r)z_j \tag{7.11}$$

or, from (7.8)

$$\log\{\pi_{is}(\mathbf{z})/\pi_{i0}(\mathbf{z})\} = \log\{\pi_{is}/\pi_{i0}\} + \sum_{j=1}^{q} \alpha_{ij}(s)z_j$$

$$(s = 1, 2, \ldots, c_i - 1). \tag{7.12}$$

When, later, we refer loosely to the "α's" we shall mean the α's of (7.11) or (7.12) which are coefficients of z's. The other parameters will be referred to, collectively, as the π's.

The interpretation of the π's is clear from their definition as "median" response probabilities. The α's may be interpreted in ways similar to those of the binary case, but now we need to consider them, separately, as functions of s and i.

The discriminating power of the latent variable z_j is indicated by the spread of the $\alpha_{ij}(s)$'s considered as functions of s. A large spread produces larger differences between the corresponding response probabilities and so a better chance of discriminating between individuals a given distance apart on the z-scale on the evidence of $\mathbf{x}_i$.

The α's are also weights in the components of (7.5). Here we are looking at the relative influence which each manifest variable has in determining the value of the component. A variable will thus be an important determinant if all of the $\alpha_{ij}(s)$'s for given i are large. That is it is the average level of the α's rather than their dispersion which counts.

The distinction between these two aspects disappears when we specialize to the binary case. The fact that $\alpha_i(0) = 0$ means that the dispersion and the mean of $\alpha_i(0)$ and $\alpha_i(1)$ are essentially the same thing.

The third way of looking at the α's is as category scores. If we think of $\alpha_{ij}(s)$ as a score associated with category s of variable i then X_j is the total score for an individual. This interpretation will be particularly relevant when we come to consider the relationship with correspondence analysis.

All of these interpretations can be given a new aspect in relation to an equivalent model based on underlying variables described in Section 7.3.

In Chapter 5 we favoured the choice of the normal function for H on the grounds that it left the likelihood unchanged under orthogonal transformation of the α's. The same advantage is gained in the polytomous case. To show this we re-define $\mathbf{A}$ to be the matrix

$$\mathbf{A} = (\boldsymbol{\alpha}_1, \boldsymbol{\alpha}_2, \ldots, \boldsymbol{\alpha}_q)$$

where $\boldsymbol{\alpha}_j' = (\alpha_{1j}(0), \alpha_{1j}(1), \ldots, \alpha_{1j}(c_1 - 1), \quad \alpha_{2j}(0), \ldots, \alpha_{2j}(c_2 - 1), \ldots, \alpha_{pj}(c_p - 1))$. If we now make an orthogonal transformation to

$$\mathbf{A}^* = \mathbf{AM}, \tag{7.13}$$

where $\mathbf{M}$ is $q \times q$ with $\mathbf{MM}' = \mathbf{I}$, then $\sum_j \alpha_{ij}(s) z_j = \mathbf{A}_i(s)\mathbf{z}$, where $\mathbf{A}_i(s)$ is the row of $\mathbf{A}$ beginning with the element $\alpha_{i1}(s)$. As before,

$$\mathbf{A}_i(s)\mathbf{z} = \mathbf{A}_i^* \mathbf{M}^{-1}\mathbf{z}$$

and the previous argument can be repeated making the transformation $\mathbf{z}^* = \mathbf{M}^{-1}\mathbf{z}$ to show that the likelihood is unchanged. This is the second indeterminacy in the model which must be removed by imposing constraints on the α's before attempting to estimate the parameters.

Nothing done so far requires or implies any ordering of the

categories. Having fitted a model we could, of course, re-label the categories so that their estimated scores were ranked in increasing or decreasing order. The analysis might then be seen as a way of uncovering an underlying metric. Conversely, if we believed the categories to be ordered *a priori* we could impose the order restrictions on the parameters and estimate the α's subject to these constraints. However, the basis for any supposed prior ordering would be likely to rest on the belief that there was some variable underlying the categories. If so it would seem more sensible to make such underlying variables explicit in the model as in the UV binary model. There are, in fact, two such models which we could use. One turns out to be equivalent to the above RF model in some circumstances but does not lead to a natural ordering of the categories. The other is the obvious generalization of the UV model for binary data. This is equivalent to an RF model but not the one given in this section. We treat each of them in turn, taking the latter first.

7.2 An underlying variable model

For the binary case we imagined that the data had arisen by recording whether or not underlying variables had values above or below fixed thresholds. The obvious generalization is to suppose that the range of each underlying variable is divided into intervals and that the manifest categorical variables record into which interval the individual falls. Thus we suppose that underlying each set of categories there is a continuous random variable whose variation is explained by the standard linear factor model

$$\boldsymbol{\xi} = \boldsymbol{\mu} + \boldsymbol{\Lambda}\mathbf{z} + \mathbf{e}. \tag{7.14}$$

On the ith dimension there is a sequence of thresholds $\tau_{i1}, \tau_{i2}, \ldots, \tau_{i,c_i-1}$ with associated indicator variables

$$\begin{aligned} x_i(s) &= 1 \quad \text{if} \quad \tau_{i,s} \leq \xi_i < \tau_{i,s+1} \quad (s = 0, 1, \ldots, c_i - 1) \\ &= 0 \quad \text{otherwise} \end{aligned}$$

where we define $\tau_{i0} = -\infty$, $\tau_{i,c_i} = \infty$. The aim is thus to make inferences about the parameters and fit of the model (7.14) on the evidence of a sample of observations on $\mathbf{x}$. We know that the underlying bivariate distributions are normal and hence we can fit the model if we can estimate the correlation coefficients. Estimates of these coefficients are known as polychoric correlations. Their computation from ordered contingency tables with more than two categories is more difficult than for 2×2 tables, but a maximum likelihood method is given in Olssen (1979) and an algorithm based on the polychoric series in Martinson and Hadman (1975). The relative advantage of assuming underlying C-type distributions is somewhat greater in the polytomous case as we shall see in the next chapter.

The model just described is not equivalent to the RF model of Section 7.1 under any circumstances unless, of course, we have only two categories on each dimension. This is easily demonstrated by defining a different RF model to which it can be made equivalent. For this purpose we define the *cumulative response function*

$$\Pi_{is}(\mathbf{z}) = \sum_{r=s}^{c_i-1} \pi_{ir}(\mathbf{z}) \quad (s = 1, 2, \ldots, c_i - 1). \tag{7.15}$$

$\Pi_{is}(\mathbf{z})$ is thus the probability of falling into category s or above of variable i for given $\mathbf{z}$. For fixed s the situation is thus exactly as it was in the binary case with categories $s, s+1, \ldots, c_i - 1$ being treated as a "positive" response. An appropriate family of models may thus be written

$$G^{-1}(\Pi_{is}(\mathbf{z})) = \alpha_{i0}(s) + \sum_{j=1}^{q} \alpha_{ij} z_j \quad (s = 1, 2, \ldots, c_i - 1; i = 1, 2, \ldots, p). \tag{7.16}$$

By the result of Section 5.3 this is equivalent to a UV model with

$$\alpha_{i0}(s) = (\tau_{i,s-1} - \mu_i)/\psi_i^{\frac{1}{2}}, \quad \alpha_{ij} = -\lambda_{ij}/\psi_i^{\frac{1}{2}}, \quad (s = 1, 2, \ldots, c_i - 1; i = 1, 2, \ldots, p) \tag{7.17}$$

and the weighted error distribution having distribution function G. The threshold parameters $\{\tau_{i,s}\}$ are thus related to the $\alpha_{i0}(s)$'s (or π's) and the factor loadings to the α's.

Because this model no longer corresponds to an RF model in the sense of the last section, we no longer have the same range of interpretations for the parameters. The parameter $\pi_{i,s}$ is still the probability that the median individual falls in category s on variable i, but the α's cannot now be interpreted as category scores or as weights in a component. This is because the posterior density of $\mathbf{z}$ is no longer a linear function of the manifest variables for any choice of G in (7.16). For example, if G is the logistic function

$$\begin{aligned}\pi_{is}(\mathbf{z}) &= \Pi_{i,s}(\mathbf{z}) - \Pi_{i,s+1}(\mathbf{z}) \\ &= \left\{1 + \exp\left(-\alpha_{i0}(s) - \sum_{j=1}^{q} \alpha_{ij}(s) z_j\right)\right\}^{-1} \\ &\quad - \left\{1 + \exp\left(-\alpha_{i0}(s+1) - \sum_{j=1}^{q} \alpha_{ij}(s+1) z_j\right)\right\}^{-1}. \end{aligned} \tag{7.18}$$

When this is substituted into (7.3) it does not yield a set of linear components.

7.3 An alternative UV model

Another way of relating categorical responses to underlying variables has been developed for modelling choice behaviour in economics (see, for example, McFadden (1982)). This is of equal relevance when one is choosing among a set of alternatives in a multiple choice question or among candidates in an election.

In this model a random variable is associated with *each* category, and the category into which an individual falls is determined by which of these underlying variables turns out to be largest. Thus, suppose that there are variables $\xi_{i0}, \xi_{i1}, \ldots, \xi_{ic_i-1}$ associated with variable i. The realized values of these variables, for any i, can be thought of as measures of the relative "attractiveness' of the categories to that individual. The largest ξ thus determines which category "wins" in the competition for the individual's vote, purchasing decision or whatever. If we postulate that a number of latent variables together with a random "error" contribute to the final attractiveness we may suppose that

$$\boldsymbol{\xi}_i = \boldsymbol{\mu}_i + \boldsymbol{\Lambda}_i \mathbf{z} + \mathbf{e}_i \quad (i = 1, 2, \ldots, p) \tag{7.19}$$

with the usual distributional assumptions about the random variables. Here $\boldsymbol{\xi}_i' = (\xi_{i0}, \xi_{i1}, \ldots, \xi_{ic_i-1})$, $\boldsymbol{\mu}_i' = (\mu_{i0}, \mu_{i1}, \ldots, \mu_{ic_i-1})$ and $\boldsymbol{\Lambda}_i = \{\lambda_{ij}(s)\}$ is a $c_i \times j$ matrix of factor loadings. We thus have a model in which a common set of latent variables (factors) account for the associations between observed categorical variables.

The parameters $\boldsymbol{\mu}_i$ may be interpreted as measuring the average attractiveness of the categories of variable i. The term $\boldsymbol{\Lambda}_i \mathbf{z}$ shows how the q latent variables influence their attractiveness and $\mathbf{e}_i$ represents the unexplained variation. All that we can observe is which of the elements in $\boldsymbol{\xi}_i$ is greatest, and using this information we have to devise methods for estimating the parameters of the model. To do this by maximum likelihood we have to specify the form of the distributions of the random variables involved. As in earlier models of the form (7.19) that we have considered, this choice is not likely to be critical since there will be a central limit effect inducing normality in the ξ's. We might therefore be content with any reasonable specification which renders the subsequent analysis tractable.

There is one such specification which makes this UV model equivalent to the RF model of Section 7.1. This requires us to assume that each residual element in (7.19) has the same Type I extreme-value distribution; that is,

$$Pr\{e_{is} \leqslant u\} = \exp(-\exp(-u)) \quad \text{for all } i \text{ and } s. \tag{7.20}$$

The significant part of this assumption is the independence of i and s, implying that the extraneous sources of variation act on the attractiveness of all categories for all variables equally.

We now demonstrate the equivalence referred to by deriving the response function $\pi_{is}(\mathbf{z})$ for this model as follows:

$$\begin{aligned}\pi_{is}(\mathbf{z}) &= Pr\{\text{individual falls into category } s \text{ of variable } i \mid z\}\\ &= Pr\{\xi_{is} \geqslant \xi_{ir}, r = 0, 1, \ldots, c_i - 1 \mid \mathbf{z}\}\\ &= \int_{-\infty}^{+\infty} Pr\{\xi_{is} \in (u, u + \mathrm{d}u)\xi_{ir} \leqslant u; r = 0, 1, \ldots, c_i - 1 \mid \mathbf{z}\}\, du\\ &= \int_{-\infty}^{+\infty} Pr\{\xi_{is} \in (u, u + \mathrm{d}u) \mid \mathbf{z}\} \prod_{\substack{r=0\\ r\neq s}}^{c_i-1} Pr\{\xi_{ir} \leqslant u \mid \mathbf{z}\} du. \qquad (7.21)\end{aligned}$$

The last step follows from the independence of the ξ's when $\mathbf{z}$ is fixed. Now

$$\begin{aligned}Pr\{\xi_{ir} \leqslant u \mid \mathbf{z}\} &= Pr\left\{\mu_{ir} + \sum_{j=1}^{q} \lambda_{ij}(r)z_j + e_{ir} \leqslant u \;\middle|\; \mathbf{z}\right\}\\ &= Pr\left\{e_{ir} \leqslant u - \mu_{ir} - \sum_{j=1}^{q} \lambda_{ij}(r)z_j \;\middle|\; \mathbf{z}\right\}\\ &= \exp\left\{-\exp(u - \mu_{ir} - \sum_{j=1}^{q} \lambda_{ij}(r)z_j\right\} \qquad (7.22)\end{aligned}$$

by (7.20). The density function is obtained by differentiation, and when this is substituted into (7.21) we obtain

$$\begin{aligned}\pi_{is}(\mathbf{z}) = &\int_{-\infty}^{+\infty} \exp\left\{-\left(u - \mu_{is} - \sum_{j=1}^{q} \lambda_{ij}(s)z_j\right)\right\}\\ &\times \exp - \sum_{r=0}^{c_i-1} \exp - \left(u - \mu_{ir} - \sum_{j=1}^{q} \lambda_{ij}(r)z_j\right) du\\ = &\exp\left(\mu_{is} + \sum_{j=1}^{q} \lambda_{ij}(s)z_j\right)\\ &\times \int_{-\infty}^{+\infty} \exp - \left\{u + e^{-u} \sum_{r=0}^{c_i-1} \exp\left(\mu_{ir} + \sum_{j=1}^{q} \lambda_{ij}(r)z_j\right)\right\} du.\end{aligned}$$

The integral may be evaluated by substituting $v = e^{-u}$, giving

$$\pi_{is}(\mathbf{z}) = \exp\left(\mu_{is} + \sum_{j=1}^{q} \lambda_{ij}(s)z_j\right) \Big/ \sum_{r=0}^{c_i-1} \exp\left(\mu_{ir} + \sum_{j=1}^{q} \lambda_{ij}(r)z_j\right)$$

$$(s = 0, 1, \ldots, c_i - 1; i = 1, 2, \ldots, p). \qquad (7.23)$$

Comparing this with (7.6), we see that the two models are identical because $\alpha_{i0}(s) \equiv \mu_{is}$ and $\alpha_{ij}(s) \equiv \lambda_{ij}(s)$.

If we wish, therefore, the parameters can be interpreted as required by the "choice" model. Whether or not this model provides a useful interpretation will depend on what the variables represent in a particular practical situation. If, for example, respondents to a questionnaire are asked which party they would vote for in an election, it seems reasonable to suppose that this would depend partly on their position on a variety of attitude scales (the z's) common to all individuals and partly on their own idiosyncracies (the e's). The same would be true for many other categorical variables arising from the expression of attitudes, preferences or opinions. The argument seems less convincing for biographical variables such as country of birth or level of education attained. However, the fact that the parameters can also be given the more neutral interpretations of Section 7.1 means that nothing vital depends on the choice.

The appearance of the extreme-value distribution in the error term where we might have expected a distribution such as the normal or logistic is unusual and not easy to interpret. The Type I distribution is positively skewed and rather like a lognormal distribution in shape. Fortunately, as noted above, its influence on the form of the distribution of ξ is not likely to be large, so it should not be critical for the form of the response function.

The identity between the α's and the λ's implies that the origin of the λ's is arbitrary for given i and j. This could be deduced directly from the UV model. Here it means that, without loss of generality, we could regard the factors $\mathbf{z}$ as having no effect on the attractiveness of one arbitrarily selected category on each manifest dimension.

An interesting feature emerges if we specialize to the binary case with $c_i = 2$ for all i. We know that our model is then equivalent to the RF model which, in turn, is equivalent to the UV model of Section 5.3. It should then follow that that UV model and the present one are equivalent in the binary case. That this is so is easily demonstrated. From (7.19),

$$\xi_{i0} - \xi_{i1} = (\mu_{i0} - \mu_{i1}) + \{\boldsymbol{\lambda}_i(0) - \boldsymbol{\lambda}_i(1)\}\mathbf{z} + (e_{i0} - e_{i1}), \quad (i = 1, 2, \ldots, p) \tag{7.24}$$

where $\boldsymbol{\lambda}_i(s) = \{\lambda_{ij}(s)\}$, $s = 0, 1$; the probability of a response in category 1, given $\mathbf{z}$, is then

$$Pr\{\xi_{i1} - \xi_{i0} \geqslant 0 \mid \mathbf{z}\} = Pr\{e_{i1} - e_{i0} \geqslant \mu_{i0} - \mu_{i1} + \{\boldsymbol{\lambda}_i(0) - \boldsymbol{\lambda}_i(1)\}\mathbf{z} \mid \mathbf{z}\}. \tag{7.25}$$

For the UV model of Chapter 5 the corresponding probability is

$$Pr\{\xi_i \geqslant \tau_i \mid \mathbf{z}\} = Pr\left\{\frac{e_i}{\psi_i^{\frac{1}{2}}} \geqslant \frac{\tau_i - \mu_i - \mathbf{\Lambda}_i\mathbf{z}}{\psi_i^{\frac{1}{2}}} \,\middle|\, \mathbf{z}\right\} \tag{7.26}$$

where $\mathbf{\Lambda}_i$ is the ith row of $\mathbf{\Lambda}$. For the two to be the same the parameter values must be chosen so that

$$\mu_{i1} - \mu_{i2} = (\tau_i - \mu_i)\psi_i^{-\frac{1}{2}} \quad \text{and} \quad \boldsymbol{\lambda}_i(1) - \boldsymbol{\lambda}_i(0) = \mathbf{\Lambda}_i\psi_i^{-\frac{1}{2}}$$

and the distribution of $c_i\psi_i^{-\frac{1}{2}}$ must be the same as that of $e_{i1} - e_{i0}$. This last requirement is easily verified, since if e_{i0} and e_{i1} have independent Type I extreme value distributions their difference has a logistic distribution with variance $\pi^2/3$. This is precisely the form which the distribution of $e_i\psi_i^{-\frac{1}{2}}$ must have if it is to be equivalent to the RF model. The two UV models thus coincide for binary data.

CHAPTER 8

Methods for Polytomous Data

8.1 Estimation by maximum likelihood

For the model of (7.4) the joint probability function of $\mathbf{x}$ is

$$f(\mathbf{x}) = \int_{-\infty}^{+\infty} \cdots \int \exp \sum_{i=1}^{p} \sum_{s=0}^{c_i-1} x_i(s) \log \pi_{is}(\mathbf{z}) h(\mathbf{z}) \, d\mathbf{z}$$

$$= \int_{-\infty}^{+\infty} \cdots \int \exp\left[\sum_{i=1}^{p} \sum_{s=0}^{c_i-1} x_i(s) \alpha_{i0}(s) + \sum_{j=1}^{q} z_j X_j - \sum_{i=1}^{p} \log \sum_{r=0}^{c_i-1} \exp\left\{ \alpha_{i0}(r) + \sum_{j=1}^{q} a_{ij}(r) z_j \right\} \right] h(\mathbf{z}) \, d\mathbf{z}. \quad (8.1)$$

As before, the log likelihood is

$$L = \sum_{h=1}^{n} \log f(\mathbf{x}_h). \quad (8.2)$$

This may be maximized with respect to the unknown parameters, but with present computing facilities this is only feasible if the number of variables and number of categories is reasonably small. Shea's (1985) POLYFAC program is limited to about 30 unknown parameters, and some examples are given in Chapter 9 which use this program. For larger problems the maximization could be effected by an E-M algorithm by an extension of the methods described for binary data. As this appears to be the most promising way forward we shall now give the necessary theory.

An E-M *Algorithm*

As in Chapter 6 we start with the conditional likelihood, for given $\mathbf{z}$. This may be written

$$L_z = \sum_{h=1}^{n} \sum_{i=1}^{p} \sum_{r=1}^{c_i-1} x_{ih}(r) \left\{ \log \frac{\pi_{ir}(\mathbf{z}_h)}{\pi_{i0}(\mathbf{z}_h)} + \log \pi_{i0}(\mathbf{z}_h) \right\}$$

$$= \sum_{h=1}^{n} \sum_{i=1}^{p} \sum_{r=1}^{c_i-1} x_{ih}(r) \left\{ \alpha_{i0}(r) + \sum_{j=1}^{q} \alpha_{ij}(r) z_{jh} + \log \pi_{i0}(\mathbf{z}_h) \right\} \quad (8.3)$$

where

$$\pi_{i0}(\mathbf{z}_h) = \left[\sum_{r=1}^{c_i-1} \exp\left\{\alpha_{i0}(r) + \sum_{j=1}^{q} \alpha_{ij}(r) z_{jh}\right\}\right]^{-1} \tag{8.4}$$

and $\mathbf{z}'_h = (z_{1h}, z_{2h}, \ldots, z_{qh})$ is the (unobserved) latent vector of the hth sample member.

The partial derivatives required for the estimating equations are then

$$\frac{\partial L_z}{\partial \alpha_{i0}(s)} = \sum_{h=1}^{n} x_{ih}(s) - \sum_{h=1}^{n} \exp\left\{\alpha_{i0}(s) + \sum_{j=1}^{q} \alpha_{ij}(s) z_{jh}\right\} \Big/ \sum_{r=0}^{c_i-1} \exp\left\{\alpha_{i0}(r) + \sum_{j=1}^{q} \alpha_{ij}(r) z_j\right\}$$

$$= \sum_{h=1}^{n} x_{ih}(s) - \sum_{h=1}^{n} \pi_{is}(\mathbf{z}_h),$$

$$(s = 1, 2, \ldots, c_i - 1;\, i = 1, 2, \ldots, p) \tag{8.5}$$

and, similarly,

$$\frac{\partial L_z}{\partial \alpha_{ij}(s)} = \sum_{h=1}^{n} x_{ih}(s) z_{jh} - \sum_{h=1}^{n} \pi_{is}(\mathbf{z}_h) z_{jh},$$

$$(s = 1, 2, \ldots, c_i - 1;\, i = 1, 2, \ldots, p;\, j = 1, 2, \ldots, q). \tag{8.6}$$

These equations should be compared with those of (6.5) which are a special case arising when $c_i = 2$ for all i. Notice that we have set $\alpha_i(0) = 0$ for all i, so these parameters do not appear in the estimating equations. For given values of $\mathbf{z}_h$ we can find estimates of the α's. For each s and i there are $q + 1$ equations, one from (8.5) and q from (8.6) to determine the parameters $\alpha_{i0}(s), \alpha_{i1}(s), \ldots, \alpha_{iq}(s)$. These are non-linear equations but can be solved iteratively using a Newton–Raphson method. This completes the M-step of the algorithm.

The E-step requires the prediction of the $\mathbf{z}$-values for each individual using the current α's by means of

$$E(\mathbf{z}_h \mid \mathbf{x}_h) = \int_{-\infty}^{+\infty} \cdots \int \mathbf{z} h(\mathbf{z}) \exp \sum_{i=1}^{p} \sum_{s=0}^{c_i-1} x_{ih}(s) \log \pi_{is}(\mathbf{z})\, \mathrm{d}\mathbf{z} / f(\mathbf{x}_h). \tag{8.7}$$

Later in this chapter, when we consider approximations to the likelihood, we shall obtain approximations to the expectation of (8.7). If these were to be used in the early cycles of the algorithm there would be a considerable saving in the amount of numerical integration to be performed.

An E-M *Algorithm Treating* **z** *as Discrete*

The method given in Chapter 6 for binary data is easily generalized to the polytomous case as follows. The partial derivatives have the same form as in (6.9), namely

$$\frac{\partial L_z}{\partial \alpha_{il}(s)} = \sum_{h=1}^{n} \frac{1}{f(\mathbf{x}_h)} \sum_{t=1}^{k} h(z_t) \frac{\partial f(\mathbf{x}_h \mid z_t)}{\partial \sigma_{il}(s)} \tag{8.8}$$

where now

$$\begin{aligned} f(\mathbf{x}_h \mid z_t) &= \prod_{i=1}^{p} \prod_{r=0}^{c_i-1} \{\pi_{ir}(z_t)\}^{x_{ih}(r)} \\ &= \prod_{i=1}^{p} \prod_{r=1}^{c_i-1} \{\pi_{ir}(z_t)\}^{x_{ih}(r)} \{\pi_{i0}(z_t)\}^{1-x_{ih}(r)}. \end{aligned} \tag{8.9}$$

The partial derivatives required in (8.8) are then obtained from

$$\begin{aligned} \frac{\partial f(\mathbf{x}_h \mid z_t)}{\partial \sigma_{il}(s)} &= f(\mathbf{x}_h \mid z_t) \frac{\partial \log f(\mathbf{x}_h \mid z_t)}{\partial \alpha_{il}(s)} \\ &= f(\mathbf{x}_h \mid z_t) \left\{ \frac{x_{ih}(s)}{\pi_{is}(z_t)} \frac{\partial \pi_{is}(z_t)}{\partial \alpha_{il}(s)} + \frac{(1 - x_{ih}(s))}{\pi_{i1}(z_t)} \frac{\partial \pi_{i1}(z_t)}{\partial \alpha_{il}(s)} \right\}. \end{aligned} \tag{8.10}$$

Substituting this expression into (8.8), reversing the order of summation and simplifying gives

$$\begin{aligned} \frac{\partial L_z}{\partial \alpha_{il}(s)} = \sum_{t=1}^{k} \Bigg[\sum_{h=1}^{n} x_{ih}(s) h(z_t \mid \mathbf{x}_h) \left\{ \pi_{i1}(z_t) \frac{\partial \pi_{is}(z_t)}{\partial \alpha_{il}(s)} - \pi_{is}(z_t) \frac{\partial \pi_{i1}(z_t)}{\partial \alpha_{il}(s)} \right\} \\ + \sum_{h=1}^{n} h(z_t \mid \mathbf{x}_h) \pi_{is}(z_t) \frac{\partial \pi_{is}(z_j)}{\partial \alpha_{il}(s)} \Bigg] \Big/ \pi_{is}(z_t) \pi_{i1}(z_t). \end{aligned} \tag{8.11}$$

So far, we have not assumed any particular form for the response function. If we use the logit version there is a great simplification because, in that case

$$\pi_{is}(z_t) = \pi_{i0}(z_t) \exp\{\alpha_{i0}(s) + \alpha_{i1}(s) z_t\}$$

and

$$\frac{\partial \pi_{is}(z_t)}{\partial \alpha_{il}(s)} = \frac{\pi_{is}(z_t)}{\pi_{i0}(z_t)} \frac{\partial \pi_{i0}(z_t)}{\partial \alpha_{il}} + z_t^m \pi_{is}(z_t) \quad (m = 0, 1). \tag{8.12}$$

Using (8.12) to eliminate the partial derivative of $\pi_{i0}(z_t)$ from (8.11), we find

$$\frac{\partial L_z}{\partial \alpha_{il}(s)} = \sum_{t=1}^{k} z_t^m \left\{ \sum_{h=1}^{n} x_{ih}(s)h(z_t \mid \mathbf{x}_h) - \pi_{is}(z_t) \sum_{h=1}^{n} h(z_t \mid \mathbf{x}_h) \right\}$$

$$= \sum_{t=1}^{k} z_t^m \{r_{it}(s) - N_t \pi_{is}(z_t)\}$$

$$(m = 0, 1; s = 1, 2, \ldots, c_i - 1; i = 1, 2, \ldots, p) \quad (8.13)$$

where

$$r_{it}(s) = \sum_{h=1}^{n} x_{ih}(s)h(z_t \mid \mathbf{x}_h), \quad N_t = \sum_{h=1}^{n} h(z_t \mid \mathbf{x}_h).$$

For a given value of s these equations are essentially the same as those of (6.15) where, again, $r_{it}(s)$ and N_t can be interpreted as pseudo-observed and expected frequencies.

At this juncture in the binary case we noted the similarity with dosage response analysis and hence that the conditional maximum likelihood estimators could be obtained by an iterative weighted least-squares analysis. The same approach could be used here.

8.2 An approximation to the likelihood

We shall now derive an approximation to the likelihood function for the logit model which can easily be maximized with respect to the parameters, thus yielding approximate estimators. It turns out that the approximations are rather poor if judged by the absolute differences. However, as we pointed out in Chapter 5, it is the profile of the set of weights which counts (i.e. their relative values) and in this respect the approximations appear to be good. This fact is illustrated by an example in Section 9.3. The result is particularly interesting in view of their equivalence to the category scores of correspondence analysis, as pointed out in Section 8.5.

The Case $q = 1$

We treat this case first in order to establish the method and then, more briefly, extend the results to cover any q. The joint probability function of $\mathbf{x}$ for the logit model is

$$f(\mathbf{x}) = \frac{1}{\sqrt{(2\pi)}} \int_{-\infty}^{+\infty} \exp\left\{ -\tfrac{1}{2}z^2 + \sum_{i=1}^{p} \sum_{r=0}^{c_i-1} x_i(r) \log \pi_{ir}(z) \right\} \mathrm{d}z. \quad (8.14)$$

Using the version of $\pi_{ir}(\mathbf{z})$ given in (7.11), (8.14) becomes

$$f(\mathbf{x}) = \left[\exp \sum_{i=1}^{p} \sum_{r=0}^{c_i-1} x_i(r) \log \pi_{is}\right]$$
$$\times \left[\frac{1}{\sqrt{(2\pi)}} \int_{-\infty}^{+\infty} \exp\left\{ -\tfrac{1}{2}z^2 + z \sum_{i=1}^{p} \sum_{r=0}^{c_i-1} x_i(r)\alpha_i(r) \right.\right.$$
$$\left.\left. - \sum_{i=1}^{p} \log \sum_{r=0}^{c_i-1} \pi_{ir} \exp \alpha_i(r)z \right\} \mathrm{d}z \right]. \tag{8.15}$$

We begin with the α's, which appear only in the second factor. The method is first to approximate this factor as described below and then to obtain estimators by maximizing the relevant part of the likelihood. We replace the last term in the exponent by its Taylor expansion taken as far as terms of the second degree in α. Thus

$$\sum_{i=1}^{p} \log \left(\sum_{r=0}^{c_i-1} \pi_{ir} \exp \alpha_i(r)z \right) \doteqdot z \sum_{i=1}^{p} \sum_{r=0}^{c_i-1} \pi_{ir}\alpha_i(r)$$
$$+ \tfrac{1}{2}z^2 \sum_{i=1}^{p} \left\{ \sum_{i=0}^{c_i-1} \pi_{ir}\alpha_i^2(r) - \sum_{r=0}^{c_i-1} \pi_{ir}\alpha_i^2(r) \right\}. \tag{8.16}$$

Recalling that the probability is unaffected by translations of the α's, there is no loss of generality if we impose the constraints

$$\sum_{r=0}^{c_i-1} \pi_{ir}\alpha_i(r) = 0, \quad (i = 1, 2, \ldots, p). \tag{8.17}$$

The right-hand side of (8.18) then simplifies to

$$\tfrac{1}{2}z^2 \sum_{i=1}^{p} \sum_{r=0}^{c_i-1} \pi_{ir}\alpha_i^2(r) = \tfrac{1}{2}z^2 \boldsymbol{\alpha}'\mathbf{P}\boldsymbol{\alpha} \tag{8.18}$$

where $\boldsymbol{\alpha}' = (\alpha_1(0), \alpha_1(1), \ldots, \alpha_1(c_1-1), \alpha_2(0), \ldots, \alpha_p(c_p-1))$ and $\mathbf{P}$ is the diagonal matrix with elements π_{ir} listed along the diagonal in dictionary order. Denoting the second factor of (8.15) by I, we now have

$$I = \frac{1}{\sqrt{(2\pi)}} \int_{-\infty}^{+\infty} \exp\{-\tfrac{1}{2}z^2 + Yz - az^2\}\, \mathrm{d}z$$
$$= (2a+1)^{-\frac{1}{2}} \exp \tfrac{1}{2}Y^2/(2a+1) \tag{8.19}$$

where $a = \frac{1}{2}\boldsymbol{\alpha}'\mathbf{P}\boldsymbol{\alpha}$ and $Y = \sum_{i=1}^{p} \sum_{r=0}^{c_i-1} x_i(r)\alpha_i(r) = \mathbf{x}'\boldsymbol{\alpha}$.

That part of the likelihood in which the α's occur may now be written

$$l = (2a+1)^{-\frac{1}{2}n} \exp \tfrac{1}{2} \sum_{h=1}^{n} Y_h^2/(2a+1) \tag{8.20}$$

where Y_h is the value of Y for the hth sample member. Now

$$\sum_{h=1}^{n} Y_h^2 = \sum_{h=1}^{n} \sum_{i=1}^{p} \sum_{j=1}^{p} \sum_{r=0}^{c_i-1} \sum_{s=0}^{c_j-1} x_{ih}(r) x_{jh}(s) \alpha_i(r) \alpha_j(s)$$
$$= \boldsymbol{\alpha}'\mathbf{X}'\mathbf{X}\boldsymbol{\alpha}. \tag{8.21}$$

The hth row of the matrix $\mathbf{X}$ consists of the set of indicator variables for the hth sample member as defined at the beginning of Chapter 7. Since each member falls into exactly one category on each variable, each row of $\mathbf{X}$ sums to p. The column sums of $\mathbf{X}$ give the numbers of sample members falling into each category on each variable. The diagonal elements of $\mathbf{X}'\mathbf{X}$ are $\left\{\sum_{h=1}^{n} x_{ih}^2(s)\right\} = \left\{\sum_{h=1}^{n} x_{ih}(s)\right\}$ and are thus the same as the column sums. The off-diagonal elements give the two-way marginal frequencies; thus

$$\sum_{h=1}^{n} x_{ih}(r) x_{jh}(s)$$

is the number of individuals in category r of variable i *and* category s of variable j. $\mathbf{X}'\mathbf{X}$ thus contains the first- and second-order marginal frequencies, and since our approximate likelihood depends on the data only through this matrix we may deduce that the estimates to be derived will also be functions of first- and second-order marginal frequencies only. Furthermore this would also be true if the exponent in the probability function were approximated by any other quadratic function of z—for example, a Taylor expansion about some point other than zero.

The relevant part of the log-likelihood may now be written

$$L = -\frac{n}{2}\log(\boldsymbol{\alpha}'\mathbf{P}\boldsymbol{\alpha} + 1) + \tfrac{1}{2}\boldsymbol{\alpha}'\mathbf{X}'\mathbf{X}\boldsymbol{\alpha}/(\boldsymbol{\alpha}'\mathbf{P}\boldsymbol{\alpha} + 1) \tag{8.22}$$

which has to be maximized subject to the constraints of (8.17). These may be written $\boldsymbol{\alpha}'\mathbf{P}^* = \mathbf{0}$ where the ith column of $\mathbf{P}^*$ has element $\pi_{i0}, \pi_{i1}, \ldots, \pi_{ic_i-1}$ in rows $(i, 0)$ to $(i, c_i - 1)$ and zeros elsewhere $(i = 1, 2, \ldots, p)$. If $\boldsymbol{\gamma}$ is a p-vector of undetermined multipliers we have to maximize

$$\phi = -\frac{n}{2}\log(\boldsymbol{\alpha}'\mathbf{P}\boldsymbol{\alpha} + 1) + \tfrac{1}{2}\boldsymbol{\alpha}'\mathbf{X}'\mathbf{X}\boldsymbol{\alpha}/(\boldsymbol{\alpha}'\mathbf{Pa} + 1) + \boldsymbol{\alpha}'\mathbf{P}^*\boldsymbol{\gamma} \tag{8.23}$$

which leads to the equations

$$-\frac{2}{n}\frac{\mathrm{d}\phi}{\mathrm{d}\boldsymbol{\alpha}}=\frac{2\mathbf{P}\boldsymbol{\alpha}}{\boldsymbol{\alpha}'\mathbf{P}\boldsymbol{\alpha}+1}-\frac{2\mathbf{X}'\mathbf{X}\boldsymbol{\alpha}}{n(\boldsymbol{\alpha}'\mathbf{P}\boldsymbol{\alpha}+1)}+\frac{2\boldsymbol{\alpha}'\mathbf{X}'\mathbf{X}\boldsymbol{\alpha}\mathbf{P}\boldsymbol{\alpha}}{n(\boldsymbol{\alpha}'\mathbf{P}\boldsymbol{\alpha}+1)^2}+\mathbf{P}^*\boldsymbol{\gamma}=0. \quad (8.24)$$

Pre-multiplying through by $\boldsymbol{\alpha}'$ and defining $\mu=\boldsymbol{\alpha}'\mathbf{X}'\mathbf{X}\boldsymbol{\alpha}/(\boldsymbol{\alpha}'\mathbf{P}\boldsymbol{\alpha}+1)$, we find that the estimates satisfy

$$\frac{\boldsymbol{\alpha}'\mathbf{P}\boldsymbol{\alpha}}{(\boldsymbol{\alpha}'\mathbf{P}\boldsymbol{\alpha}+1)}-\frac{\mu}{n}+\frac{\mu\boldsymbol{\alpha}'\mathbf{P}\boldsymbol{\alpha}}{n(\boldsymbol{\alpha}'\mathbf{P}\boldsymbol{\alpha}+1)}=0$$

or

$$\mu/n=\boldsymbol{\alpha}'\mathbf{P}\boldsymbol{\alpha}. \quad (8.25)$$

Substituting in (8.24) then yields

$$\mathbf{P}\boldsymbol{\alpha}-\frac{\mathbf{X}'\mathbf{X}\boldsymbol{\alpha}}{(n+\mu)}+\tfrac{1}{2}\mathbf{P}^*\boldsymbol{\gamma}=0$$

or

$$\mathbf{P}^{-1}\mathbf{X}'\mathbf{X}\boldsymbol{\alpha}=(n+\mu)\{\boldsymbol{\alpha}+\tfrac{1}{2}\mathbf{P}^{-1}\mathbf{P}^*\boldsymbol{\gamma}\}. \quad (8.26)$$

We next express $\boldsymbol{\gamma}$ in terms of $\boldsymbol{\alpha}$ and thus eliminate $\boldsymbol{\gamma}$ from (8.26) to obtain an equation for $\boldsymbol{\alpha}$ alone. Pre-multiplying both sides of (8.26) by $(\mathbf{P}^*)'$ and re-calling that $\boldsymbol{\alpha}'\mathbf{P}^*=\mathbf{0}$, we have

$$(\mathbf{P}^*)'\mathbf{P}^{-1}\mathbf{X}'\mathbf{X}\boldsymbol{\alpha}=\tfrac{1}{2}(n+\mu)(\mathbf{P}^*)'\mathbf{P}^{-1}\mathbf{P}^*\boldsymbol{\gamma}=\tfrac{1}{2}(n+\mu)\boldsymbol{\gamma}. \quad (8.27)$$

Substituting for $\boldsymbol{\gamma}$ in (8.26) gives

$$\mathbf{P}^{-1}[\mathbf{I}-\mathbf{P}^*(\mathbf{P}^*)'\mathbf{P}^{-1}\}\mathbf{X}'\mathbf{X}\boldsymbol{\alpha}=(n+\mu)\boldsymbol{\alpha}. \quad (8.28)$$

It thus appears that $\boldsymbol{\alpha}$ is an eigenvector of the matrix pre-multiplying it on the left-hand side of (8.28); $(n+\mu)$ will be the corresponding eigenvalue. Since an eigenvector is determined only up to an arbitrary scale factor, (8.25) is necessary to render the solution unique.

To complete the solution we need the corresponding equations for the π's. A first approximation can be obtained by expanding the log-likelihood as far as terms of the first degree in $\boldsymbol{\alpha}$. The second factor in (8.15) involves no term of this order, so we may obtain our estimates by maximizing the first factor. Imposing the constraints $\sum_{r=0}^{c_i-1}\pi_{ir}=1$ for all i, we find the unrestricted maximum of

$$\phi=\sum_{h=1}^{n}\sum_{i=1}^{p}\sum_{r=0}^{c_i-1}x_{ih}(r)\log\pi_{ir}+\sum_{i=1}^{p}\theta_i\sum_{r=0}^{c_i-1}\pi_{ir}. \quad (8.29)$$

This is the standard multinomial estimation problem with solution

$$\hat{\pi}_{is} = \sum_{h=1}^{n} x_{ih}(s)/n \quad (i = 1, 2, \ldots, p; s = 0, 1, \ldots, c_i - 1)$$

or

$$\hat{\boldsymbol{\pi}} = \mathbf{X}'\mathbf{1}/n \tag{8.30}$$

where $\mathbf{1}$ is an n-vector of 1's and $\boldsymbol{\pi}$ is the vector of the π_{is}'s in dictionary order. The π's are thus estimated by the corresponding marginal proportions.

The final step in finding estimators of $\boldsymbol{\alpha}$ is to replace $\mathbf{P}$ and $\mathbf{P}^*$ in (8.28) by their estimates derived from (8.30). However, (8.28) can first be simplified by noting that

$$(\mathbf{P}^*)'\mathbf{P}^{-1} = \begin{pmatrix} \mathbf{1}' & \mathbf{0}' \ldots \mathbf{0}' \\ \mathbf{0}' & \mathbf{1}' \ldots \mathbf{0}' \\ \mathbf{0}' & \mathbf{0}' \ldots \mathbf{1}' \end{pmatrix},$$

a matrix with p rows. It then follows that

$$(\mathbf{P}^*)'\mathbf{P}^{-1}\mathbf{X}' = \begin{pmatrix} \mathbf{1}' \\ \mathbf{1}' \\ \vdots \\ \mathbf{1}' \end{pmatrix}$$

and hence that

$$(\mathbf{P}^*)'\mathbf{P}^{-1}\mathbf{X}'\mathbf{X}\boldsymbol{\alpha} = \frac{1}{n}\begin{pmatrix} \hat{\boldsymbol{\pi}}' \\ \boldsymbol{\pi}' \\ \boldsymbol{\pi}' \end{pmatrix}\boldsymbol{\alpha}$$

by (8.30). Now since $(\mathbf{P}^*)'\boldsymbol{\alpha} = \mathbf{0}$ it follows that $\boldsymbol{\pi}'\boldsymbol{\alpha} = 0$. When we replace $\boldsymbol{\pi}$ throughout by $\hat{\boldsymbol{\pi}}$, the second term on the left-hand side of (8.28) vanishes and we are left with

$$\mathbf{P}^{-1}\mathbf{X}'\mathbf{X}\boldsymbol{\alpha} = (n + \mu)\boldsymbol{\alpha} \tag{8.31}$$

as the eigen equation for $\boldsymbol{\alpha}$.

Let $\boldsymbol{\alpha}_0$ be a normalized eigenvector of $\hat{\mathbf{P}}^{-1}\mathbf{X}'\mathbf{X}$, then any other vector $\boldsymbol{\alpha} = K\boldsymbol{\alpha}_0$ will also satisfy (8.31). But is must also satisfy (8.25), which implies that

$$K^2\boldsymbol{\alpha}_0'\hat{\mathbf{P}}\boldsymbol{\alpha}_0 = \mu/n.$$

Let θ be the eigenvalue associated with $\boldsymbol{\alpha}_0$; then, from (8.31),

$\theta = n + \mu$, so

$$K^2 \boldsymbol{\alpha}_0' \hat{\mathbf{P}} \boldsymbol{\alpha}_0 = \frac{\theta}{n} - 1$$

or

$$K^2 = (\theta - n)/n\boldsymbol{\alpha}_0' \hat{\mathbf{P}} \boldsymbol{\alpha}_0. \tag{8.32}$$

The ambiguity in the sign of K arises from the fact that changing the sign of all the α's does not change the likelihood.

It remains to determine which of the eigen solutions maximizes the likelihood. Substituting back into (8.22) gives

$$L = -\frac{n}{2}\left(\log\frac{\theta}{n} - \frac{\theta}{n} - 1\right). \tag{8.33}$$

We thus only have a solution if $\theta > 0$ and, in this range, L is monotonic increasing in θ/n. The log-likelihood is maximized by taking the largest eigenvalue and its associated eigenvector.

The approximate solution is thus obtained by first finding the normalized eigenvector of the matrix $\hat{\mathbf{P}}^{-1}\mathbf{X}'\mathbf{X}$ and then multiplying it by the scaling factor K given by (8.32).

For purposes of computation it is much easier to extract the eigenvalues and vectors of a symmetric positive definite matrix. The matrix $\hat{\mathbf{P}}^{-1}\mathbf{X}'\mathbf{X}$ is not symmetric positive definite, but if we introduce new parameters, $\boldsymbol{\beta}$, defined by $\boldsymbol{\beta} = \hat{\mathbf{P}}^{\frac{1}{2}}\boldsymbol{\alpha}$, then (8.31) becomes

$$\hat{\mathbf{P}}^{-1}\mathbf{X}'\mathbf{X}\hat{\mathbf{P}}^{-\frac{1}{2}}\boldsymbol{\beta} = (n + \mu)\hat{\mathbf{P}}^{-\frac{1}{2}}\boldsymbol{\beta}$$

or

$$\hat{\mathbf{P}}^{-\frac{1}{2}}\mathbf{X}'\mathbf{X}\hat{\mathbf{P}}^{-\frac{1}{2}}\boldsymbol{\beta} = (n + \mu)\boldsymbol{\beta}. \tag{8.34}$$

The matrix on the left-hand side is now symmetric positive definite and the estimate of $\boldsymbol{\beta}$ can be obtained exactly as described for $\boldsymbol{\alpha}$. The final step is to revert to the α's using

$$\hat{\boldsymbol{\alpha}} = \hat{\mathbf{P}}^{-\frac{1}{2}}\hat{\boldsymbol{\beta}}. \tag{8.35}$$

The Case $q \geqslant 2$

The treatment of the general case follows the same lines, so we shall treat it more briefly by concentrating on the points at which it differs from the case $q = 1$. The second factor in (8.15) becomes

$$(2\pi)^{-q/2} \int_{-\infty}^{+\infty} \cdots \int \exp\left[-\sum_{j=1}^{q} z_j^2 - \sum_{j=1}^{q} z_j Y_j - \sum_{i=1}^{p} \log \sum_{r=0}^{c_i-1} \pi_{ir} \right.$$
$$\left. \times \exp\left\{ -\sum_{j=1}^{q} \alpha_{ij}(r) z_j \right\} \right] d\mathbf{z}. \tag{8.36}$$

The approximation now consists of replacing the last term in the exponent by a quadratic form derived from the Taylor expansion, giving

$$\sum_{i=1}^{p} \log \sum_{r=0}^{c_i-1} \pi_{ir} \exp\left\{ -\sum_{j=1}^{q} \alpha_{ij}(r) z_j \right\} \doteqdot \mathbf{z}'\mathbf{B}\mathbf{z} + \mathbf{b}'\mathbf{z} \tag{8.37}$$

where $\mathbf{b}$ is a q-vector with

$$b_j = -\sum_{i=1}^{p} \sum_{r=0}^{c_i-1} \pi_{ir}\alpha_{ij}(r) \quad (j = 1, 2, \ldots, q) \tag{8.38}$$

and $\mathbf{B}$ is a $q \times q$ matrix with elements b_{jk} given by

$$2b_{jk} = \sum_{i=1}^{p} \left\{ \sum_{r=0}^{c_i-1} \pi_{ir}\alpha_{ij}(r)\alpha_{ik}(r) - \sum_{r=0}^{c_i-1} \pi_{ir}\alpha_{ij}(r) \sum_{r=0}^{c_i-1} \pi_{ir}\alpha_{ik}(r) \right\}. \tag{8.39}$$

If, as before, we centre the α's for each variable at zero, then

$$\sum_{r=0}^{c_i-1} \pi_{ir}\alpha_{ij}(r) = 0 \quad (i = 1, 2, \ldots, p)$$

and

$$\mathbf{b} = \mathbf{0} \quad \text{and} \quad 2\mathbf{B} = \mathbf{A}'\mathbf{P}\mathbf{A}$$

where $\mathbf{A}' = (\boldsymbol{\alpha}_1, \boldsymbol{\alpha}_2, \ldots, \boldsymbol{\alpha}_q)$, $\boldsymbol{\alpha}_q$ being the vector of α's for the jth latent variable.

The exponent in (8.36) can thus be written as

$$-\tfrac{1}{2}\{\mathbf{z}'\mathbf{z} + 2\mathbf{Y}'\mathbf{z} + \mathbf{z}'\mathbf{A}'\mathbf{P}\mathbf{A}\mathbf{z}\}$$

where $\mathbf{Y}' = (Y_1, Y_2, \ldots, Y_q)$. Completing the square, the integral reduces to

$$|\mathbf{I} + \mathbf{A}'\mathbf{P}\mathbf{A}|^{-\frac{1}{2}} \exp \tfrac{1}{2}\mathbf{Y}'(\mathbf{A}'\mathbf{P}\mathbf{A} + \mathbf{I})^{-1}\mathbf{Y}. \tag{8.40}$$

Noting that $\mathbf{Y} = \mathbf{A}'\mathbf{x}$ and that

$$\mathbf{Y}'(\mathbf{A}'\mathbf{P}\mathbf{A} + \mathbf{I})\mathbf{Y} = \operatorname{tr}\{(\mathbf{A}'\mathbf{P}\mathbf{A} + \mathbf{I})^{-1}\mathbf{Y}\mathbf{Y}'\}$$

the relevant part of the log-likelihood becomes

$$L = -\frac{n}{2}\log|\mathbf{A}'\mathbf{P}\mathbf{A} + \mathbf{I}| + \tfrac{1}{2}\operatorname{tr}\{(\mathbf{A}'\mathbf{P}\mathbf{A} + \mathbf{I})^{-1}\mathbf{A}'\mathbf{X}'\mathbf{X}\mathbf{A}\}. \tag{8.41}$$

Our approximate estimators are obtained by maximizing this expression with respect to $\mathbf{A}$. However, a new feature is that there is no unique maximum, for if we make an orthogonal transformation to $\mathbf{A}^* = \mathbf{A}\mathbf{N}$, $(\mathbf{N}'\mathbf{N} = \mathbf{I})$, the value of L is unchanged. The proof of this result depends on the two following well-known theorems:

(a) the determinant of a product of two square matrices equals the product of the determinants;
(b) $\operatorname{tr} \mathbf{N}'\mathbf{Z}\mathbf{N} = \operatorname{tr} \mathbf{Z}$ where $\mathbf{N}$ is orthogonal.

Taking the two parts of (8.41) in turn,

$$\begin{aligned}\log|(\mathbf{A}^*)'\mathbf{P}\mathbf{A}^* + \mathbf{I}| &= \log|\mathbf{N}'\mathbf{A}'\mathbf{P}\mathbf{A}\mathbf{N} + \mathbf{I}| = \log|\mathbf{N}'(\mathbf{A}'\mathbf{P}\mathbf{A} + \mathbf{I})\mathbf{N}| \\ &= \log|\mathbf{N}'\mathbf{N}| + \log|\mathbf{A}'\mathbf{P}\mathbf{A} + \mathbf{I}| = \log|\mathbf{A}'\mathbf{P}\mathbf{A} + \mathbf{I}|\end{aligned}$$

and

$$\begin{aligned}\operatorname{tr}[\{(\mathbf{A}^*)'\mathbf{P}\mathbf{A}^* + \mathbf{I}\}^{-1}(\mathbf{A}^*)'\mathbf{X}'\mathbf{X}\mathbf{A}^*] \\ &= \operatorname{tr}\{(\mathbf{N}'\mathbf{A}'\mathbf{P}\mathbf{A}\mathbf{N} + \mathbf{I})^{-1}\mathbf{N}'\mathbf{A}'\mathbf{X}'\mathbf{X}\mathbf{A}\mathbf{N}\} \\ &= \operatorname{tr}[\{\mathbf{N}'(\mathbf{A}'\mathbf{P}\mathbf{A} + \mathbf{I})^{-1}\mathbf{N}\}^{-1}\mathbf{N}'\mathbf{A}'\mathbf{X}'\mathbf{X}\mathbf{A}\mathbf{N}] \\ &= \operatorname{tr}[\mathbf{N}'(\mathbf{A}'\mathbf{P}\mathbf{A} + \mathbf{I})^{-1}\mathbf{N}\mathbf{N}'\mathbf{A}'\mathbf{X}'\mathbf{X}\mathbf{A}\mathbf{N}] \\ &= \operatorname{tr}\{(\mathbf{A}'\mathbf{P}\mathbf{A} + \mathbf{I})^{-1}\mathbf{A}'\mathbf{X}'\mathbf{X}\mathbf{A}\}.\end{aligned}$$

To obtain a solution we must therefore impose sufficient constraints to render the solution unique; other solutions can then be obtained by orthogonal rotation.

Since $\mathbf{A}'\mathbf{P}\mathbf{A}$ is a symmetric positive definite matrix, it can be written in the form $\mathbf{M}\boldsymbol{\Delta}\mathbf{M}'$ where $\boldsymbol{\Delta}$ is a diagonal matrix of its eigenvalues and $\mathbf{M}$ is an orthogonal matrix composed on the eigenvectors. It follows that

$$\boldsymbol{\Delta} = (\mathbf{A}\mathbf{M})'\mathbf{P}(\mathbf{A}\mathbf{M})$$

and therefore that the matrix $\mathbf{A}'\mathbf{P}\mathbf{A}$ can be diagonalized by the transformation $\mathbf{A}^* = \mathbf{A}\mathbf{M}$. We may thus obtain the unique maximum of L when $\mathbf{A}'\mathbf{P}\mathbf{A}$ is diagonal and derive all others by orthogonal transformation. Now since the determinant of a diagonal matrix is the product of its diagonal elements, (8.41) may now be written

$$L = \sum_{j=1}^{q}\left[-\frac{n}{2}\log\{\boldsymbol{\alpha}_j'\mathbf{P}\boldsymbol{\alpha}_j + 1\} + \frac{\boldsymbol{\alpha}_j'\mathbf{X}'\mathbf{X}\boldsymbol{\alpha}_j}{1 + \boldsymbol{\alpha}_j'\mathbf{P}\boldsymbol{\alpha}_j}\right]. \tag{8.42}$$

Note that the jth diagonal element of $\mathbf{A}'\mathbf{P}\mathbf{A}$ is only equal to $\boldsymbol{\alpha}_j'\mathbf{P}\boldsymbol{\alpha}_j$ when the former is diagonal, as we are assuming. If (8.42) is now compared with (8.22) it will be seen that each term in the sum is of the same form as in the previous case. L in (8.42) is thus maximized if each term individually is maximized and this is a problem we have already solved. It is thus clear that the vectors $\boldsymbol{\alpha}_j$ will all be eigenvectors of $\mathbf{P}^{-1}\mathbf{X}'\mathbf{X}$. Let $\boldsymbol{\alpha}_1, \boldsymbol{\alpha}_2, \ldots, \boldsymbol{\alpha}_q$ be any q such eigenvectors; then according to (8.33),

$$L = -\frac{n}{2}\sum_{j=1}^{q}\left(\log\frac{\theta_j}{n} - \frac{\theta_j}{n} + 1\right). \tag{8.43}$$

It follows that $\theta_1, \theta_2, \ldots, \theta_j$ should be the largest j eigenvalues and that we may extract factors as long as $\theta_j \geqslant n$. This restriction imposes a limit on the number of factors that can be extracted.

An interesting property of this solution is that, to the same degree of approximation, the components are uncorrelated. Thus the dispersion matrix of $\mathbf{Y}$ is

$$D(\mathbf{Y}) = \mathbf{A}'E(\mathbf{xx}' - E\mathbf{x}E\mathbf{x}')\mathbf{A}.$$

Conditioning on $\mathbf{z}$, $E\mathbf{x} = E\boldsymbol{\pi}(\mathbf{z})$ and $E\mathbf{xx}' = E\boldsymbol{\pi}(\mathbf{z})\boldsymbol{\pi}'(\mathbf{z})$, where $\boldsymbol{\pi}'(\mathbf{z}) = (\pi_{i0}(\mathbf{z}), \ldots, \pi_{ic_i-1}(\mathbf{z}), \ldots, \pi_{pc_p-1}(\mathbf{z}))$. Expanding the (i, r)th element of this vector as far as terms of the first degree in the α's,

$$\begin{aligned}\pi_{ir}(\mathbf{z}) &= \pi_{ir}\left[1 - \sum_{j=1}^{q} z_j\left\{\alpha_{ij}(r) - \sum_{r=0}^{c_i-1} \pi_{ir}\alpha_{ij}(r)\right\}\right] \\ &= \pi_{ir}\left\{1 - \sum_{j=1}^{q} z_j\alpha_{ij}(r)\right\}.\end{aligned}$$

Hence

$$\pi_{ir}(\mathbf{z}) - E\pi_{ir}(\mathbf{z}) = -\pi_{ir}\sum_{j=1}^{q} z_j\alpha_{ij}(r)$$

and

$$\operatorname{cov}\{\pi_{ir}(\mathbf{z}), \pi_{ks}(\mathbf{z})\} = \pi_{ir}\pi_{ks}\sum_{j=1}^{q}\alpha_{ij}(r)\alpha_{kj}(s).$$

Expressed in matrix notation,

$$D(\mathbf{x}) = \mathbf{PA}(\mathbf{PA})'$$

and therefore,

$$D(\mathbf{Y}) = \mathbf{A}'\mathbf{PA}(\mathbf{PA})'\mathbf{A} = (\mathbf{A}'\mathbf{PA})^2. \tag{8.44}$$

It thus follows that if we have constrained $\mathbf{A}'\mathbf{PA}$ to be diagonal then, to a first approximation, the components will be uncorrelated.

The estimation of the π's requires no new theory and we thus proceed as before, using the estimators of (8.30).

8.3 Binary data as a special case

We now have to justify the results given in Section 6.4, and this is done by putting $c_i = 2$, for all i, in the formulae of this chapter. The vector $\boldsymbol{\alpha}$ is then given by

$$\boldsymbol{\alpha}' = (\alpha_1(0), \alpha_1(1), \alpha_2(0), \alpha_2(1), \ldots, \alpha_p(0), \alpha_p(1)).$$

The elements of the vector have been constrained so that

$$\pi_{i0}\alpha_i(0) + \pi_{i1}\alpha_i(1) = 0 \quad (i = 1, 2, \ldots, p). \tag{8.45}$$

In the binary case we estimated a p-vector of parameters, also called $\boldsymbol{\alpha}$, given by $\boldsymbol{\alpha}' = (\alpha_1, \alpha_2, \ldots, \alpha_p)$; α_i was the score associated with category 1 of each variable—the score for the first category was zero. Since both approaches are essentially estimating the *difference* between the scores for each variable, the relationship between the two sets of α's is that

$$\alpha_i = \alpha_i(1) - \alpha_i(0),$$

or, using (8.45), that

$$\alpha_i(0) = -\pi_{i1}\alpha_i \quad \text{and} \quad \alpha_i(1) = \pi_{i0}\alpha_i, \quad (i = 1, 2, \ldots, p).$$

Since, in the binary case, $\pi_{i1} = \pi_i$ and $\pi_{i0} = 1 - \pi_i$ we may write

$$\boldsymbol{\alpha} = \begin{pmatrix} -\pi_1 & 0 & \ldots & 0 \\ 1-\pi_1 & 0 & \ldots & \vdots \\ 0 & -\pi_2 & \ldots & \vdots \\ 0 & 1-\pi_2 & \ldots & \vdots \\ \vdots & 0 & \ldots & \vdots \\ \vdots & \vdots & & -\pi_p \\ 0 & 0 & \ldots & 1-\pi_p \end{pmatrix} \boldsymbol{\alpha}^* = \mathbf{S}\boldsymbol{\alpha}^*, \text{ say} \qquad (8.46)$$

where, temporarily, $(\boldsymbol{\alpha}^*)' = (\alpha_1, \alpha_2, \ldots, \alpha_p)$. Substituting (8.46) into (8.31) yields

$$\hat{\mathbf{P}}^{-1}\mathbf{X}'\mathbf{X}\mathbf{S}\boldsymbol{\alpha}^* = (n + \mu)\mathbf{S}\boldsymbol{\alpha}^*. \qquad (8.47)$$

We convert this into another eigen equation in two steps. First pre-multiply both sides by $\mathbf{S}'\hat{\mathbf{P}}$, giving

$$\mathbf{S}'\mathbf{X}'\mathbf{X}\mathbf{S}\boldsymbol{\alpha}^* = (n + \mu)\mathbf{S}'\hat{\mathbf{P}}\mathbf{S}\boldsymbol{\alpha}^*. \qquad (8.48)$$

Next pre-multiply by $(\mathbf{S}'\hat{\mathbf{P}}\mathbf{S})^{-1}$, then

$$(\mathbf{S}'\hat{\mathbf{P}}\mathbf{S})^{-1}\mathbf{S}'\mathbf{X}'\mathbf{X}\mathbf{S}\boldsymbol{\alpha}^* = (n + \mu)\boldsymbol{\alpha}^*. \qquad (8.49)$$

This is the required equation and it remains to simplify the left-hand side. As before, we replace the π's by their estimators given by (8.30). Direct multiplication then gives

$$\hat{\mathbf{S}}'\hat{\mathbf{P}}\hat{\mathbf{S}} = \text{diag}\{\hat{\pi}_i(1 - \hat{\pi}_i)\} = \hat{\mathbf{Q}}, \quad \text{say}$$

and

$$\hat{\mathbf{S}}'\mathbf{X}'\mathbf{X}\hat{\mathbf{S}} = \sum_{h=1}^{n} (\mathbf{x}_h - \hat{\boldsymbol{\pi}})'(\mathbf{x}_h - \hat{\boldsymbol{\pi}}) = n\hat{\boldsymbol{\Omega}}, \quad \text{say}$$

where $\mathbf{x}'_h = (x_{1h}, x_{2h}, \ldots, x_{ph}) = (x_{1h}(1), x_{2h}(1), \ldots, x_{ph}(1))$. Equation (8.49) may thus be written

$$\hat{\mathbf{Q}}^{-1} n \hat{\mathbf{\Omega}} \boldsymbol{\alpha}^* = (n + \mu)\boldsymbol{\alpha}^* \quad (8.50)$$

which is identical in form with (8.31) and can therefore be solved by exactly the same method. If we make the further transformation $\boldsymbol{\beta} = \mathbf{Q}^{\frac{1}{2}}\boldsymbol{\alpha}^*$ as in (8.34), the equation becomes

$$\hat{\mathbf{Q}}^{-\frac{1}{2}} \hat{\mathbf{\Omega}} \hat{\mathbf{Q}}^{-\frac{1}{2}} \boldsymbol{\beta} = \left(1 - \frac{\mu}{n}\right)\boldsymbol{\beta} \quad (8.51)$$

and the left-hand matrix will be recognized as the "phi-coefficient" matrix for the binary variables. It follows that the β's are the principal components derived from the phi-coefficient matrix. The α's are thus not principal component weights because of the weightings $\{\hat{\pi}_i(1 - \hat{\pi}_i)\}^{-\frac{1}{2}}$ arising from multiplication by $\mathbf{Q}^{-\frac{1}{2}}$. Since the α's derived by this means are also the scores that would be derived from a multiple correspondence analysis, this result also shows the connection of that technique with principal components.

8.4 Methods for the UV model

There are several methods by which a UV model of the kind introduced in Section 7.2 can be fitted to data. Full maximum likelihood is a possibility but would rapidly increase in complexity with the number of categories. However, there are two other approaches which are feasible in the present state of knowledge.

Reduction to Binary Case

If the α's were known, the π's could be estimated by a straightforward extension of the method in the binary case. Thus if n_{is} denotes the number of sample members in category s of variable i, the π's can be estimated from the equations

$$E\pi_{is}(\mathbf{z}) = n_{is}/n, \quad (i = 1, 2, \ldots, p; s = 0, 1, \ldots, c_i - 1). \quad (8.52)$$

As (7.18) shows, the left-hand side involves the two unknowns $\pi_{i,s+1}$ and $\pi_{i,s}$; but since $\pi_{i,c_i} = 0$ we can begin with $s = c_i - 1$ which enables us to estimate π_{i,c_i-1} and then work down the list. An equivalent way is to estimate the cumulative π's deriving from the binary form of the model indicated by (7.15) and then obtain the required estimates by differencing. The α's are estimated by reducing each of the variables to a dichotomy and then using one of the methods for binary data. If the sample members are concentrated in a small number of categories and if the dichotomy is made as near to the point which produces equal frequencies in each part as possible, the loss of information will be relatively modest. A refinement would be to make the split at different points and combine the estimates by some kind of averaging, but this has not been investigated. This approach is described more fully in Bartholomew (1980).

Correlation Methods

These methods involve factor analysing the estimated correlation coefficients of the underlying bivariate distributions. If these distributions are normal, the tetrachoric correlations of the bivariate case are replaced by polychoric correlations as noted in Chapter 7. The α's are then estimated in the same way as in the binary case and the π's as described above.

Methods based on underlying C-type distributions are somewhat easier to handle. In particular the cross-product ratio parameter, γ, can easily be estimated by conditional maximum likelihood as follows. The frequencies in the cells of the contingency table for any pair of variables will be multinomially distributed. For a C-type distribution the cell probabilities will be functions of the marginal distributions and of γ only. If we consider a cell bounded by the threshold values as

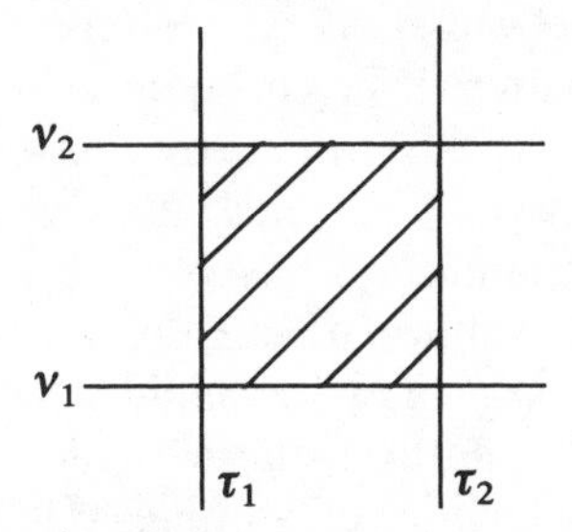

illustrated, the cell probability is

$$F(\tau_2, \nu_2) - F(\tau_2, \nu_1) - F(\tau_1, \nu_2) + F(\tau_1, \nu_1)$$

where F is the bivariate distribution function. The cells at the boundaries of the table are dealt with by making the appropriate thresholds infinite. Now

$$F(\tau, \nu) = [T - \{T^2 - 4\gamma(\gamma - 1)F(\tau)F(\nu)\}]/\{2(\gamma - 1)\} \quad (8.53)$$

where $T = 1 + (\gamma - 1)\{F(\tau) + F(\nu)\}$. If we replace $F(\tau)$ and $F(\nu)$ by their estimates derived from the marginal proportions, the likelihood becomes a function of the single parameter γ only, and it can be maximized by routine numerical methods (see Fachel (1986)). We then have the same choice of correlation coefficients as in the binary case.

This approach is easy to use and sets no practical limit on the number of factors or variables which can be accommodated. Its drawbacks are that it can only be used if the UV model of Section 7.2 is appropriate, and it provides no information on goodness of fit or standard errors of the estimates. Experience with other methods suggests that these may well be large.

8.5 Further observations

Methods for polytomous data are more recent and less well-tried than those for binary data and this is an area where more research is needed. However, limited experience suggests that some features of those methods apply here also. Using full maximum likelihood for the RF method, it is possible to estimate the asymptotic standard errors of the estimators and some examples are given in Chapter 9. These tend to be large, especially for the α's and we shall see that they increase rapidly as α increases beyond a certain limit. There seems to be a greater tendency for estimators to diverge, and the propensity for this to happen appears to depend, as before, on the sample size and on the number of variables. We have also noted that trouble is more likely to arise with a poorly fitting model. Testing the goodness of fit again poses problems if the number of parameters is large relative to the number of response patterns. A chi-squared or likelihood ratio test will only be satisfactory if, after any grouping needed to give sufficiently large expected frequencies, there is a positive number of degrees of freedom remaining.

It has been implicit in much of our discussion that the number of variables is relatively large and the number of categories relatively small. This would be the norm in those applications where the object is to construct scales of abilities or attitudes. However, it is possible to use the model when the number of categories is large and the number of variables is small. In particular we could use it to analyse two-way contingency tables with many categories.

Insofar as the RF model can be viewed as assigning scores to the categories of the manifest variables it has a marked similarity to correspondence analysis (see, for example, Greenacre (1984)). Both approaches can be regarded either as methods of scoring categories or of scoring individuals. In both cases the individual score (the component value in the present case) is the sum of the scores of the categories on each variable into which the individual falls. It would not, therefore, be surprising to find a deeper connection between the two types of analysis. It turns out that, in essence, they differ only in the function which is optimized. Viewed from one angle, correspondence analysis assigns scores to individuals such that the variation between categories on each dimension is maximized. This is achieved by maximizing the average "between-groups sum of squares" subject to a fixed total sum of squares. In the latent variable approach it is the likelihood ratio for testing the hypothesis that the α's are equal on each dimension against a general alternative which is maximized. This ratio may be regarded as a measure of inequality among the α's, since when it is maximized we have made the differences as "significant" as possible.

The similarity between the two methods is, in fact, even closer because the approximate maximum likelihood estimators obtained in this chapter are essentially the same as the correspondence analysis scores. The RF model therefore provides a model-based version of correspondence analysis with the benefits of a framework for inference. Conversely the simpler computational properties of correspondence analysis suggest its use as a first approximation to the latent variable approach. These are matters requiring further investigation.

CHAPTER 9

Examples of the Analysis of Categorical Data

9.1 Scope and purpose

Much of the theory described in Chapters 5 to 8 is relatively new and depends for its application on computer software which is not yet widely available. In order to demonstrate the potential of these methods we shall illustrate their use on a variety of sets of data. Some of these have appeared frequently as examples in earlier studies and their use here enables us to make comparisons with other methods. The remainder are new and have been selected from a wide range of applications, mainly in the social sciences. Some of the examples raise interesting substantive questions but this is not the place to pursue them.

The models fitted are the RF models for binary data as given in Section 5.2 and for polytomous data as given in Section 7.1. For binary data this also includes UV models since, as we showed, they are equivalent to RF models. It does, however, exclude the UV models for polytomous data of the kind discussed in Section 7.2. For those models the recommended way to proceed is to estimate correlation coefficients of the polychoric or Chambers form and then to estimate the parameters by factor analysis methods. These are then interpreted as in the binary case, and the methodology has already been covered and illustrated in Chapter 3.

Two further restrictions have been imposed. Apart from a brief summary in Section 9.2, all the examples involve between 3 and 5 variables. With polytomous data it is difficult to have many more variables with the software currently available, but in the binary case up to 50 or 60 variables can be accommodated. For illustrative purposes the limitation to a small number of variables makes for a more compact presentation and enables all the essential features to be covered.

The second restriction is to one-factor models. This is partly dictated by what is currently possible within the RF framework but also reflects the considerable practical importance of this case. In an exploratory study with a large number of variables we recommend that the dimensionality of the factor space be first explored using a UV model.

Because of the lack of theory on goodness of fit the choice of the number of factors must be rather subjective, but bootstrap or cross-validation investigations may help to clarify matters. When it comes to constructing ordinal scales for well-established latent variables, as is almost exclusively the case in latent trait applications, then the one-factor model is what is required.

It cannot be emphasized too often that very large samples are required to estimate the many parameters involved with adequate precision. We shall see that this is especially true for polytomous data.

9.2 Analysis of binary data using the logit model

Cancer Knowledge

This set of data, which has been analysed many times before, exemplifies many of the features which arise in the analysis of a small number of binary variables with a large sample size. It comes from a study by Lombard and Doering (1947) on knowledge of cancer. As well as obtaining information about cancer knowledge, questions were asked about the respondents' source of knowledge in general. This was elicited by asking whether or not the following were sources of general information:

1) Radio 2) Newspapers 3) Solid reading 4) Lectures

A positive response was coded 1 and negative response zero. Fitting a latent variable model might be proposed as a way of constructing a measure of how well-informed a person is about matters in general. A second stage might then be to see how knowledge about cancer relates to such general knowledge. We thus need to know whether there is evidence of a single latent dimension underlying the four binary variables and, if so, how to construct scores for individuals which locate them on that underlying dimension.

There are $2^4 = 16$ possible response patterns which are listed in Table 9.1, along with their observed frequencies and various other quantitites which we shall need. The response patterns are listed, not in the natural order, but in increasing order of their estimated ranking on the latent variable. This greatly facilitates the interpretation of the last three columns.

The number of degrees of freedom for p variables is $2^p - 2p - 1$ which has been reduced by one in this case because two categories have been amalgamated. Parameter estimates are given in Table 9.2.

All calculations were made with the FACONE program, but the same results would be obtained using the NAG routine G11SAF. The original calculations here and elsewhere were made to a greater

Table 9.1 The fit and scores obtained by fitting the logit/probit model to Lombard and Doering's data

Response pattern	Frequence	Expected frequency	Total score	Component score	"y-score"
0000	477	467.4	0	.000	.208
1000	63	70.8	1	.721	.280
0001	12	16.6	1	.766	.284
0010	150	155.9	1	1.340	.343
1001	7	3.1	2	1.487	.358
1010	32	33.3	2	2.061	.416
0011	11	8.0	2	2.107	.421
1011	4	2.0	3	2.828	.497
0100	231	240.5	1	3.401	.558
1100	94	82.2	2	4.121	.639
0101	13	20.3	2	4.167	.644
0110	378	362.3	2	4.741	.710
1101	12	8.5	3	4.888	.727
1110	169	181.6	3	5.462	.792
0111	45	46.0	3	5.507	.797
1111	31	30.5	4	6.228	.870
Total	1729	1729.0			

$\chi^2 = 11.68$ with 6 degrees of freedom (2 categories with small expected frequencies have been amalgamated).

accuracy and have been rounded to an accuracy sufficient for practical use.

The fit displayed in Table 9.1 is generally good, though the value of the log-likelihood statistic judged as a chi-squared random variable is not far short of the 5% significance level. This provides reasonable grounds for believing that the association among the variables is largely if not entirely explained by their common dependence on a single latent variable. From Table 9.2 we see that all of the α_{i1}'s are positive, confirming our expectation that the latent variable measures what all have in common, namely the transmission of information. We are therefore justified in regarding the latent variable as a measure of how well-informed a person is.

Table 9.2 Parameter estimates and asymptotic standard errors for the logit/probit model fitted to Lombard and Doering's data by maximum likelihood

Variable (i)	$\hat{\alpha}_{i1}$	s.e. ($\hat{\alpha}_{i1}$)	$\hat{\alpha}_{i0}$	$\hat{\pi}_i$	s.e. ($\hat{\pi}_i$)
1	.72	.09	−1.29	.22	.01
2	3.40	1.14	.60	.65	.05
3	1.34	.17	−.14	.47	.02
4	.77	.15	−2.71	.06	.01

The component scores (see (5.6)) provide a ranking of response patterns on the latent dimension. The "y-scores" do likewise but also give the expected percentile of the distribution of the latent variable at which an individual with the given response pattern stands (see p. 101).

The total score is simply the number of positive responses, and a comparison of this with the component scores shows what has been gained by using the model instead of relying on crude addition of responses. It is the different weighting applied to the x's which accounts for the difference and, in particular, the large weight given to newspapers (x_2). In fact anyone who reads a newspaper ranks higher on the latent scale than anyone who does not. A person whose only source of information is newspapers ranks above one who uses all other sources except newspapers. The second most important variable is solid reading (x_3) and it is noteworthy that both media involving the written word rank higher than those relying on aural communication.

We notice that the standard errors are relatively large in spite of the sample size, and especially so in the case of x_2. This is typical and is found whenever an α is larger than, say 2. However, in this example a fairly large deviation from 3.40 would be needed to alter seriously the conclusions reached from Table 9.1.

The estimates of the π's are much more precise. These show that newspapers (x_2) and solid reading (x_3) are also the most common sources of information, with lectures (x_4) being a minority interest. The alternative versions of these parameters given by $\alpha_{i0} = \text{logit}\ \pi_i$ merely express the same information in a different but less immediately useful form.

It is interesting to see how dependent the estimates are on the choice of the functions G and H (see (5.2)) and on the method of fitting. Estimates of the π's are virtually the same, and some results for the α's are given in Table 9.3.

The fits of the three models using maximum likelihood are almost identical, and the variation in the estimates is not sufficient to alter the

Table 9.3 Comparison of various estimates of the α's for Lombard and Doering's data

	Maximum likelihood			Methods of Section 6.3			Methods of Section 6.2	
i	logit/ probit	logit/ logit	probit/ probit	I	II	III	Maximum likelihood	Principal factor
1	.72	.73	.76	.86	.83	.91	.82	.82
2	3.40	4.20	3.46	2.08	2.17	2.37	2.26	2.19
3	1.34	1.39	1.47	1.46	1.37	1.45	1.43	1.36
4	.77	.76	.69	1.01	.93	1.20	.82	.94

conclusions one would draw from the analysis. As might be expected, the largest discrepancy occurs with $\hat{\alpha}_{21}$ but even here the difference is less than the estimated standard error. The various approximate methods agree well among themselves and agree with maximum likelihood in the relative importance which they assign to the four items. The largest difference from maximum likelihood is in the estimation of α_{21} which is the least well-determined under any method. These discrepancies are typical of those which occur when the α's are highly variable, but even here they would not lead to significant differences in interpretation.

The program also provides asymptotic estimates of the dispersion matrix of the estimators. The only correlations which seem to be consistently large are those between an α and the corresponding π. For the present example these are, in order, $-.43$, $.81$, $-.12$, $-.67$. The sign is positive if $\pi > \frac{1}{2}$ and negative otherwise, since an increase in α tends to move the associated $\hat{\pi}$ towards the nearer end-point of the interval $(0, 1)$. Large correlations arise if either $\hat{\alpha}$ is large (as with $\hat{\alpha}_{21}$ here) or if $\hat{\pi}$ is close to the end of the range (as with $\hat{\pi}_4$). The correlation matrix of the $\hat{\alpha}$'s is

$$\begin{bmatrix} 1 & -.22 & .13 & .12 \\ & 1 & -.73 & -.25 \\ & & 1 & .27 \\ & & & 1 \end{bmatrix}$$

Again we note that the only large correlation involves the poorly determined $\hat{\alpha}_{21}$.

It may be added that the number of iterations is liable to be much larger when one of the α's is large.

The Law School Admission Test, Section VI

This is another data set which has been re-analysed many times and hence where it is possible to compare different methods. Unlike the previous example, it is based on a test designed to measure a single latent ability scale. The main interest is therefore in whether the attempt to construct items which are indicators solely of this ability has been successful and, if so, what the parameter estimates tell us about the items.

The test consisted of 5 items taken by 1000 individuals. The fit of the logit/probit model by maximum likelihood yielded a goodness of fit measure $\chi^2 = 15.30$ on 13 degrees of freedom, which indicates a very satisfactory fit. We may conclude that the items are all indicators of a single latent variable. The parameter estimates and their standard errors are given in Table 9.4.

Table 9.4 Parameter estimates and asymptotic standard errors for the logit/probit model fitted to the LSAT Section VI data by maximum likelihood

Item (i)	$\hat{\alpha}_{i1}$	s.e. ($\hat{\alpha}_{i1}$)	$\hat{\pi}_i$	s.e. ($\hat{\pi}_i$)	Binomial s.e.
1	.83	.26	.94	.01	.01
2	.72	.19	.73	.02	.01
3	.89	.23	.56	.02	.02
4	.69	.19	.78	.02	.01
5	.66	.21	.89	.01	.01

The α-estimates show that the items have very similar discriminating powers, and the differences between them are of the same order as the standard errors. The π's show a range of difficulties, item 1 being the easiest and item 3 the most difficult. The standard errors are all small, and we have also included in the table binomial standard errors given by $\{\hat{\pi}_i(1-\hat{\pi}_i)/n\}^{\frac{1}{2}}$. These would be exact if the α's were zero and provide a reasonable approximation (though an underestimate) in cases like this where the α's are roughly equal.

The full set of data with the various scalings is set out in Table 9.5. They are arranged in blocks according to the total score.

In this case we notice that the component score offers very little more than the total score. The component score does discriminate between response patterns in the total score groups, but, given the size of the standard errors, one would have little confidence in the significance of such fine distinctions. If the values in the last column are plotted, clear breaks appear between the groups. With this test, therefore, one could reasonably dispense with the α's and scale individuals on the basis of their total score. This does not mean, of course, that the model is useless in this case. Without it we would have had no basis for concluding that the effect of weighting the scores would be negligible.

Table 9.6 compares estimates obtained by various other methods with those given above. Muthén's GLS method (generalized least squares) is that described in Section 6.1; the ULS method (unweighted least squares) is equivalent to factor analysing the tetrachoric coefficients rather than the Chambers coefficients of Section 6.2. The "minres" estimates were obtained by using Method II on the Chambers coefficient with $v = 0.74$. This is only a selection of possible estimates, but it is sufficient to show that, for most practical purposes, there is no essential difference. The variation which does not occur is small when compared with the standard errors. Experience suggests that this is usually the case if the α's are of similar magnitude. For small numbers of variables the full maximum likelihood method is to be preferred because it also yields standard errors and a test of

Table 9.5 The fit and scores obtained by fitting the logit/probit model to the Law School Admission Test Section VI data

Response pattern	Frequency	Expected frequency	Total score	Component score
00000	3	2.3	0	0
00001	6	5.9	1	.66
00010	2	2.6	1	.69
01000	1	1.8	1	.72
10000	10	9.5	1	.83
00100	1	.7	1	.89
00011	11	8.9	2	1.35
01001	8	6.4	2	1.38
01010	0	2.9	2	1.41
10001	29	34.6	2	1.48
10010	14	15.6	2	1.51
00101	1	2.6	2	1.55
11000	16	11.3	2	1.55
00110	3	1.2	2	1.58
01100	0	.9	2	1.61
10100	3	4.7	2	1.71
01011	16	13.6	3	2.07
10011	81	76.6	3	2.17
11001	56	56.1	3	2.21
00111	4	6.0	3	2.24
11010	21	25.7	3	2.24
01101	3	4.4	3	2.27
01110	2	2.0	3	2.30
10101	28	25.0	3	2.37
10110	15	11.5	3	2.40
11100	11	8.4	3	2.44
11011	173	173.3	4	2.89
01111	15	13.9	4	2.96
10111	80	83.5	4	3.06
11101	61	62.5	4	3.10
11110	28	29.1	4	3.13
11111	298	296.7	5	3.78
Total	1000	1000.0	—	—

$\chi^2 = 15.30$ on 13 degrees of freedom.

goodness of fit. But if p is large, methods such as Method II of Section 6.3 applied to the cross-product ratios or the Chambers coefficients are very much faster and almost as good. Asymptotic standard errors are available for the GLS method (see Muthén (1978), Table 1) and these are very close to those of the maximum likelihood estimators.

Table 9.6 Estimates of α's for the Law School Admission Test Section VI data by various methods

i	logit/probit	Muthén GLS	Method of Section 6.3 I	Method of Section 6.3 II	Muthén ULS	Method of Section 6.2 Minres
1	.83	.76	.83	.84	.73	.84
2	.72	.83	.78	.90	.81	.79
3	.89	.93	.94	1.00	1.00	.94
4	.69	.83	.73	.84	.73	.71
5	.66	.69	.68	.76	.63	.64

Arithmetic Reasoning Test

Mislevy (1985) gives frequency distributions of response patterns for samples of American youth on the Armed Services Vocational Aptitude Battery. The individuals were classified by sex and colour. The results given here relate to black and white females. (The males are excluded here because the fit of the one-factor model was less good.) For black females, $\chi^2 = 6.42$ on 3 degrees of freedom ($P \doteqdot .10$) and for white females, $\chi^2 = 8.39$ on 6 degrees of freedom ($P \doteqdot .21$). In both cases we may reasonably infer that the observed associations are consistent with a single latent variable and hence that the battery of four items are indicators of arithmetic reasoning ability. However, Table 9.7 shows striking differences in the parameter estimates of the logit/probit model.

The result for white females is exactly what one would expect with an ability test. The items show a range of difficulty, with none too easy and none too hard, and approximately of equal discriminating power. The scaling given by the component score in Table 9.8 is virtually the same as that of the total score. By contrast, with the black females, item 1 plays a dominating role. Although it is very imprecisely determined it is large enough to divide the sample into two entirely

Table 9.7 Parameter estimates and standard errors for test data on black and white females (standard errors in brackets)

	Black		White	
i	$\hat{\alpha}_{i1}$	$\hat{\pi}_i$	$\hat{\alpha}_{i1}$	$\hat{\pi}_i$
1	14.39(*)	.56(*)	1.04(.32)	.64(.04)
2	.38(.23)	.42(.04)	1.24(.39)	.64(.04)
3	.37(.24)	.28(.04)	1.00(.30)	.49(.04)
4	.19(.24)	.25(.04)	1.44(.45)	.38(.05)

* The standard errors estimated in the usual way are so large as to be untrustworthy.

Table 9.8 Comparison of score distributions for black and white females

Black				White			
Response pattern	Frequency	Total score	Component score	Response pattern	Frequency	Total score	Component score
0000	29	0	.00	0000	20	0	0
0001	8	1	.19	0001	14	1	1.00
0010	7	1	.37	1000	23	1	1.04
0100	14	1	.38	0100	20	1	1.24
0011	3	2	.56	0001	8	1	1.44
0101	5	2	.57	1010	9	2	2.04
0110	6	2	.75	0110	11	2	2.24
0111	0	3	.94	1100	18	2	2.28
1000	14	1	14.39	0011	2	2	2.44
1001	10	2	14.58	1001	8	2	2.48
1010	11	2	14.76	0101	5	2	2.68
1100	19	2	14.77	1110	20	3	3.28
1011	2	3	14.96	1011	6	3	3.49
1101	5	3	14.96	0111	7	3	3.68
1110	8	3	15.14	1101	15	3	3.72
1111	4	4	15.33	1111	42	4	4.72
Total	145				228		

separate groups. All those responding positively to this item rank above those who do not. Within these two groups, individuals are ranked according to their total score on the remaining three items. The difference between the two groups is less marked for the π's (the difficulty parameters). The relative difficulties are similar for the two groups, but the median black has a lower probability of giving a positive response than the median white.

This example poses interesting questions of interpretation which cannot be fully explored here. If the latent ability which is being measured is the same for both groups then we could conclude (a) that the median of the white ability distribution is higher than the black and (b) that whereas all items contribute equally to determining a white's position on the scale, item 1 plays a dominating role for blacks. Presumably, therefore, there would be cultural or other differences which affect the way in which the items reveal the latent ability in the two groups. On the other hand it might be argued that the two latent abilities are not the same, meaning that the same measuring instrument measures different things when applied to different groups. What is measured is then the result of interaction between the instrument and the subject. It is difficult to see how the dilemma could be resolved statistically, but the statistical analysis does sharply focus what it is that has to be explained.

Some Other Binary Examples

(1) *Examples from Goodman (1978)* In Chapter 2 we noted that Goodman had fitted a variety of latent class models to binary response data, and one example based on Coleman's panel data was given in Table 2.1. There we found an excellent fit using a pair of cross-classified binary latent variables. In the light of this, it would be surprising if a model with a single continuous latent variable provided an adequate fit. This expectation is confirmed by finding $\chi^2 = 239.3$ on 7 degrees of freedom for the logit/probit model. However, two of the other examples given by Goodman are fitted very well by the logit/probit model, as the following results show:

Lazarsfeld–Stouffer Questionnaire: $\chi^2 = 7.4$ on 7 degrees of freedom.

Stouffer–Toby Questionnaire: $\chi^2 = 5.9$ on 3 degrees of freedom.

Goodman (1978) fitted models based on what he called a quasi-independence concept in which those response patterns not corresponding to a Guttman scale were random. In both cases he obtained excellent fits, indicating again how difficult it may be to distinguish between competing models.

(2) *Examples with larger values of p* In many studies of attitudes and abilities up to 50 or 60 items may be used. There is no problem of principle in fitting the RF models in this case, though it may be expensive in computer time. Nevertheless, compared with the cost of collecting the data it is still likely to be minimal. Even with 10 items there are 1024 response patterns and this makes a detailed presentation of the results difficult. Scheussler's scales of social life feelings (Scheussler (1982)) typically involve this number of items, and the logit model has been fitted to give scales of such things as "trustworthiness" and "feeling low". On sample sizes of about 500 the α's will typically lie in the interval $(1, 2)$ with standard errors of about 0.2. All items therefore carry similar weight in forming the scales, indicating that the items have been well chosen. With such a large number of response patterns a goodness of fit test is not possible but comparisons of the observed and fitted first- and second-order margins suggest very good fits.

Aitkin, Anderson and Hinde's (1981) modelling of the teaching styles data sought to identify latent classes, but they also remarked that one might assume a formal/informal continuous latent scale. A fit of the logit/probit model supports that conclusion by producing a component with positive loadings on those items where a positive response indicates a traditional (or formal) approach and loads negatively on the remainder. The loadings vary considerably in absolute value from 0.1 to 2 with standard errors between 0.1 for the smaller ones to 0.3 for the larger. The latter are rather smaller than those for Scheussler's data on very similar sample sizes, reflecting a tendency for an increase in p to lead to more precise estimation. The fit of the one-factor model appears to be less good here and this suggests that a second factor may be needed. The results of fitting such a model were reported in Bartholomew's contribution to the discussion of Aitkin *et al.* (1981).

The most extensive application of RF models to date is to a large set of educational testing data collected by the National Foundation for Educational Research at Slough. It gives results of tests of mathematical ability of children for about 50 items. There were 26 samples each of about 500 children. The full results of the analysis were included in an end-of-grant research report to the Economic and Social Research Council and are now deposited in the National Lending Library. The logit model was fitted, so the figures quoted here have been scaled up by the factor $\pi/\sqrt{3}$ to make them comparable with estimates quoted from other examples. A typical loading would be around 1.3 with a standard error of about 0.2 with extremes in the neighbourhood of .5 and 3.5. These are too widely scattered to support the hypothesis of equal discriminating power but close enough for the "total correct" score to give a good indication of ranking on the ability scale. The items show a range of difficulty spanning the whole interval (0, 1) and the observed and fitted first- and second-order margins show no evidence of the need for a second factor. We may conclude that the test instruments were well suited to their purpose.

In Chapter 3 we discussed Heywood cases which arise in factor analysis when an estimate of a ψ_i approaches zero. A similar phenomenon arises with RF models where it manifests itself in estimates of an α diverging to infinity. This is mainly a small sample problem and for a given sample size it is less likely to occur if p is large or if the α's are of similar size. The advice given in Section 3.6 applies here and need not be repeated. The advantages of having a large p may be illustrated by the educational testing data discussed above. The estimation procedure converged in all 26 cases with the full set of variables. The model was then fitted to the first 32, 17 and 14 variables. In the 14-variable case the loading on the second variable began to diverge. The loadings on the first 4 variables are given in Table 9.9; those for the 14-variable fit were those achieved after 100 iterations. This table also illustrates how the estimated loadings for any item may depend on which other items are included in the test. This is similar to the familiar situation in regression analysis where changes can occur in regression coefficients if variables are added or

Table 9.9 Estimates of α's obtained from the first *m* items for a sample of size 517 with 53 items

m	14	17	32	53
$\hat{\alpha}_{11}$	1.68	1.43	1.00	.86
$\hat{\alpha}_{21}$	10.01	1.95	1.09	.89
$\hat{\alpha}_{31}$	4.46	1.62	.99	.85
$\hat{\alpha}_{41}$	.89	1.18	1.27	1.19

deleted from the set. It is not surprising that this should be so since all items are correlated and those left out will exert an influence through their correlation with those left in. It does, however, caution us against regarding the α's as measuring intrinsic properties of the items.

9.3 Analysis of polytomous data using the logit model

Staff Assessment Data

The first example comes from the results of a staff assessment exercise in a large public organization. Each of 405 managers were assessed on 13 aspects of their work using a 5-point ordered scale. Three of the variables are used in this illustration. The higher categories were rarely used, so some amalgamation was carried out. On variable 1, categories 4 and 5 were amalgamated and on variables 2 and 3 categories, 4, 5 and 6. The resulting data constitute a $4 \times 3 \times 3$ contingency table. The frequencies are set out in Table 9.10, where, again, they are listed in order of estimated component score. By fitting a one-factor RF model we aim to see whether the three categorical variables can be replaced by a single underlying latent variable and, if so, how each of the observed variables contributes to the latent scale.

The calculations were made with the POLYFAC program and all entries have been rounded for ease of presentation. In order to economize on space, results are only presented for response patterns which have a non-zero observed frequency. In this case the missing cases are 002, 102, 202, 300, 301, 302, 310 and 320. The component scores for these patterns can easily be obtained using the parameter estimates in Table 9.11.

It is clear that the one-factor model provides an excellent fit, so the observed response patterns are entirely consistent with a single underlying dimension which one might describe as success in the job.

Further elucidation of the relationship between the latent variable and the observed variables can be had by looking at the parameter estimates given in Table 9.11. We notice first that the standard errors of the α's are very large, so if we interpret the α's as category scores

Table 9.10 The fit and scores obtained by fitting the logit/probit model to staff assessment data

Response pattern	Frequency	Expected frequency	Component score	"y-score"	"z-score"
000	1	4.3	0.00	.03	−2.16
100	7	6.8	.95	.05	−1.85
200	1	1.3	1.73	.08	−1.61
001	2	3.9	1.95	.09	−1.55
010	3	3.1	2.36	.11	−1.42
101	13	11.0	2.90	.14	−1.25
020	1	.9	2.94	.14	−1.24
110	10	9.8	3.31	.17	−1.12
201	5	3.5	3.68	.19	−1.00
120	5	3.6	3.89	.21	−.93
210	3	3.4	4.08	.23	−.86
011	12	11.8	4.31	.25	−.78
220	1	1.5	4.66	.29	−.66
021	4	5.3	4.88	.31	−.58
111	64	69.6	5.26	.35	−.44
121	36	37.9	5.84	.42	−.23
211	38	41.5	6.03	.45	−.16
221	31	26.7	6.61	.52	.05
012	1	2.1	6.78	.54	.12
311	4	3.3	6.92	.56	.17
022	1	1.6	7.36	.61	.34
321	3	2.6	7.50	.63	.39
112	37	28.8	7.73	.66	.48
122	23	27.0	8.31	.72	.71
212	34	35.6	8.51	.75	.79
222	41	40.2	9.09	.81	1.05
312	5	6.8	9.40	.84	1.20
322	11	9.8	9.98	.89	1.49
Others	0	1.4	—	—	—
Total	405	405.1			

$\chi^2 = 2.876$ on $19 - 14 - 1 = 4$ degrees of freedom (after amalgamating into 19 groups to avoid small expected frequencies).

they must be regarded as very imprecisely determined on a sample of this size. The categories were ordered, and it is reassuring to note that, for each variable, the α's form a monotonic sequence. In Section 7.1 (p. 134) we noted several ways in which the α's could be interpreted. Viewing them as weights in the formation of the components, we see that Variable 3 carries most weight, and this accounts for the preponderance of response patterns in the highest category of this variable at the upper end of the scale of Table 9.10. Similarly Variable 2 is more important than Variable 1. Interpreted as discrimination parameters, it appears that Variable 3, with the largest spread of

Table 9.11 Maximum likelihood estimates and standard errors for the logit/probit model fitted to staff assessment data

Variable (i)	Category (s)	$\hat{\alpha}_i(s)$	s.e. $\hat{\alpha}_i(s)$	Approx $\alpha_i(s)$	$\hat{\pi}_{is}$	s.e. ($\hat{\pi}_{is}$)
1	0	0	—	0	.05	—
	1	.95	.38	.76	.52	.03
	2	1.73	.47	1.33	.40	.03
	3	2.61	.75	1.79	.03	.02
2	0	0	—	0	.02	—
	1	2.36	1.25	1.22	.59	.03
	2	2.94	1.28	1.68	.39	.03
3	0	0	—	0	.02	—
	1	1.95	.87	1.00	.70	.10
	2	4.43	1.87	1.68	.28	.09

Note: The program does not give $\hat{\pi}_{i0}$ which can easily be obtained by subtraction.

scores, is the best single variable for distinguishing the overall quality of individuals. Looking at the α's as category scores, the categories of Variables 1 and 3 are roughly equally spaced, whereas for Variable 2 categories 2 and 3 are very close together and might be amalgamated with little loss of information.

The π's are much more precisely estimated. Recall that π_{is} is the probability of the "median individual" falling into category s on variable i. As one might have anticipated, such an individual will most likely be placed in the middle category(ies).

Table 9.11 also gives the approximate maximum likelihood estimates for the α's obtained by the method described in Section 8.2; they are also used as starting values for the iterative procedure used to obtain the estimates. They are equivalent to the scores obtained by multiple correspondence analysis and therefore of independent interest. At first sight the approximation is very poor; the true estimates exceed the starting values by a factor of roughly 2 and this seems to be a typical result. However, as we observed earlier, it is the profile of the scores which is important for interpretation. In every case the approximate scores place the categories in the same order and at roughly the same spacing. (It is this fact which makes them good starting values for iteration, because it appears to be important to have the right ordering to get rapid convergence.) One way of judging the accuracy of the approximate estimates is to compute the approximate components and to see how the rankings assigned to response patterns compare. If this is done for data of this example, we find that the ranking assigned to response patterns using the final and initial estimates in constructing the components has correlation coefficient

0.97. In this sense therefore the approximate estimates give results which differ little from the exact estimates.

Life Satisfaction Data

In Chapter 2 we showed that a latent class model with three classes provided an excellent fit to data from Clogg (1979) on life satisfaction. We suggested that the classes might represent differing levels of satisfaction with life in general. However, the fact that the questions had invited respondents to classify their answers into three categories—low, medium and high—may have led to a tendency for individuals to think of their satisfaction with things in general in the same categorical terms. This might have induced a grouping on the lines suggested by the model. It might seem more plausible to postulate a continuum of general satisfaction, and this leads us to investigate the fit of the RF model. The parameter estimates are given in the first part of Table 9.12, together with the goodness of fit test. The latter provides moderate, though not strong, evidence for rejecting the model.

There is also an anomaly in the α's for Variable 1. Instead of increasing monotonically as we move from low to high (as happens with Variables 2 and 3), the value for "medium" is actually lower than that for "low". However, a closer examination of the results shows that it would be premature to reject the model out of hand. We notice that $\hat{\alpha}_1(0)$ and $\hat{\alpha}_1(1)$ are very close compared with their standard errors, so that the inversion in the expected ordering has no real significance. Question 1 (on family) also differs from the others in that there is marked tendency for people to have higher satisfaction on this

Table 9.12 Parameter estimates of the RF model of Section 7.1 for Clogg's life satisfaction data

Variable (i)	Category (s)	$\hat{\alpha}_{i1}(s)$	s.e. $\hat{\alpha}_{i1}(s)$	$\hat{\pi}_{is}$	$\hat{\alpha}_{i1}(s)$	s.e. $\hat{\alpha}_{i1}(s)$	$\hat{\pi}_{is}$
1	0	0	—	.04	{ 0	—	.15
	1	−.08	.27	.11			
	2	1.56	.29	.85	1.62	.24	.85
2	0	0	—	.12	0	—	.12
	1	.41	.17	.29	.42	.17	.29
	2	1.54	.22	.59	1.55	.22	.59
3	0	0	—	.12	0	—	.12
	1	.37	.16	.37	.37	.16	.37
	2	1.49	.21	.51	1.49	.21	.51

$\chi^2 = 27.28$ on 14 degrees of freedom. $\chi^2 = 7.299$ on 7 degrees of freedom. The standard errors of the $\hat{\pi}$'s are all 0.02 to two decimals.

item, as the estimates of the π's show. The median respondent has a higher probability (0.85) of being highly satisfied on this item than on either of the others. This means that the two lower categories are used less often, and since they are more distant in the perceptions of most people the distinction between them may be blurred. We have therefore amalgamated the low and medium categories on Variable 1 and refitted the model. The estimates are given in the second part of Table 9.12. On Variables 2 and 3 there is little change and the estimates of the π's are identical. On Variable 1, $\hat{\alpha}_1(2)$ is almost the same. The major change is in the overall goodness of fit which is now very good. We may thus conclude that the pattern of response to these three items is consistent with their being dependent on a single latent dimension. Since the α's are all positive, the component scores appear to measure satisfaction in general. The spread of the α's on each item is very similar, indicating that the variables have similar discriminating power (except for "low" and "medium" on Variable 1 as already noted). In all cases a move from the medium to high categories represents a bigger shift in general satisfaction than from low to medium.

Table 9.13 The fit of the RF model of Section 7.1 to Clogg's life satisfaction data. (On Variable 1, 0 denotes "low or medium")

Response pattern	Frequency	Expected frequency	Component score	Latent class assigned in Chapter 2
000	31	24.8	0	I
001	37	43.3	.37	I
010	26	34.7	.42	II
011	70	65.3	.79	II
002	19	18.0	1.49	I
020	23	21.5	1.55	II
100	23	22.8	1.62	II
012	36	33.8	1.91	II
021	52	49.4	1.92	II
101	49	52.8	1.99	II
110	45	43.9	2.04	II
111	117	109.3	2.41	II
022	43	46.3	3.05	III
102	54	51.2	3.11	III
120	64	64.2	3.17	III
112	126	133.0	3.53	III
121	191	195.8	3.54	III
122	466	461.9	4.66	III
Total	1472	1472.0		

It is interesting to compare the estimated ranking on the latent scale with the allocation to latent classes given on p. 34. This is done in Table 9.13 for the case when the low and medium categories for Variable 1 have been grouped. We have broken the table at two points suggested by the occurrence of gaps on the component score scale. (A case could be argued for regarding those who respond 122 as a separate class.) We notice from the last column that this closely accords with the results for the latent class analysis. For practical purposes, therefore, the two models are in substantial agreement. However, the fact that the RF model gives a scaling of both individuals and categories may be used to argue that it is both more plausible and useful. Masters (1985) reached somewhat similar conclusions using a version of the Rasch model.

Employment in Small Industry

It often happens that a category such as "don't know" is included among a set of possible answers to a question in a survey. For example, in a simple case there might be three answers: yes, no and don't know.

If we were to contemplate using a UV model we would have to decide on an ordering of the responses and, in particular, whether "don't know" should be the middle category. One advantage of the RF model is that the data is allowed to speak for itself by determining the category scores. Our final example illustrates this using data from Leimu (1983).

The study was of a sample of 469 employees from small industry in Finland obtained in 1975. There are three variables defined by the following questions:

(1) Was there any alternative choice of job when coming to your present job? (0 = no, 1 = don't know, 2 = yes)
(2) Is the job permanent? (0 = very unsure or quite unsure, 1 = don't know, 2 = quite sure or very sure)
(3) Were you unemployed in the last three years? (0 = no, 1 = yes)

The results of fitting the one-factor RF model are given in Tables 9.14 and 9.15 in the usual format.

Although the standard errors of the α's are large, their ordering across the categories strongly suggests that the "don't know" response is correctly placed in the middle position on each variable. The π's are more precisely estimated. Had the "yes" response on Variable 3 been coded 0, the $\hat{\alpha}$'s would have had the same signs on all variables and, in retrospect, this would have been more natural. It is clear from the goodness of fit test that the response patterns are satisfactorily

Table 9.14 Parameter estimates of the RF model of Section 7.1 for Leimu's employment data

Variable (i)	Category (s)	$\hat{\alpha}_i(s)$	s.e.	$\hat{\pi}_{is}$	s.e.
1	0	0	—	.28	—
	1	−.30	.30	.22	.02
	2	−.62	.27	.50	.03
2	0	0	—	.12	—
	1	−.44	.46	.19	.02
	2	−1.67	.61	.69	.03
3	0	0	—	.88	—
	1	1.13	.56	.12	.04

explained by a single latent dimension. The component score may thus be thought of as an index which ranks employees according to the ease with which they can find secure employment. A person who ranks high will be someone who has been unemployed, had little choice of job in the past and no secure future in the present job. A person with the complementary attributes will be at the opposite end of the scale.

Table 9.15 Observed and expected frequencies and component scores for Leimu's employment data

Response pattern	Frequency	Expected frequency	Component score
220	145	145.3	−1.78
120	54	57.4	−1.46
020	72	67.5	−1.17
210	33	33.6	−1.06
110	22	15.7	−.74
221	17	16.6	−.65
200	24	23.1	−.62
010	14	21.6	−.44
121	9	8.5	−.33
100	9	11.9	−.30
021	11	12.7	−.04
000	21	17.9	.00
211	7	6.9	.07
111	6	4.1	.39
201	6	6.6	.51
011	7	7.1	.69
101	2	4.4	.83
001	10	8.2	1.13
Total	469	469.1	

$\chi^2 = 7.55$ on 6 degrees of freedom.

This example provides a good illustration of how the model can achieve a reduction in dimensionality of social survey data. The POLYFAC program also produces scores on the assumptions that the latent variable has a normal (z-scores) or uniform (y-scores) prior distribution and these, especially the normal scores, may be useful if they are to be subjected to further analysis.

Computer Programs

The three main statistical packages, BMDP, SAS and SPSS, all provide methods appropriate for the linear factor model. The SPSS-X version of the latter is much to be preferred to its predecessors. A useful comparison of the facilities available in the three systems is given by Tabachnik and Fidell (1983), but this was too early to include the SPSS-X version. A comparison of the performance of the three packages as they existed at the time is given in MacCullum (1984).

Two other packages, known as LISREL and COSAN, deal with a more general problem of which factor analysis is a special case, but these would not normally be used for routine factor analysis.

Programs for fitting a one-factor RF model with binary or polytomous manifest variables by maximum likelihood, known, respectively, as FACONE and POLYFAC, were written by Dr Brian Shea at the London School of Economics. They have been made available on a limited basis to educational users. An improved version of FACONE is publicly available in the NAG Library (G11SAF and G11SBF).*

The method of fitting the probit/probit model proposed by Bock and Aitkin (1981) is included in the program BILOG distributed by Scientific Software, Inc., P.O. Box 536, Moorsville, Indiana 46158. TESTFACT, available from the same source, allows the factor analysis of categorical data using correlations, including the tetrachoric coefficient.

Programs for some of the simpler methods, such as latent class analysis, can easily be written using standard languages and facilities.

* Numerical Algorithms Group (1987). Library Manual, Mk. 12. NAG Central Office, 256 Banbury Road, Oxford, OX2 7DE.

Bibliography

This bibliography contains the references in the text, together with a selection of other publications. The latter include some of the major early contributions, now mainly of historical interest, and a number of other publications chosen because of their importance or relevance for the approach to the field followed in the book.

Aitkin, M. (1980) Contribution to discussion in D. J. Bartholomew (1980) "Factor analysis for categorical data". *J. Roy. Statist. Soc.*, **B, 42,** 312–14.

Aitkin, M., Anderson, D. and Hinde, J. (1981) "Statistical modelling of data on teaching styles". *J. Roy. Statist. Soc.,* **A, 144,** 419–61.

Akaike, H. (1983) "Information measures and model selection". *Bull. Int. Statist. Inst.,* **50,** Book I, 277–90.

Albert, A. and Anderson, J. A. (1984) "On the existence of maximum likelihood estimates in logistic regression models". *Biometrika,* **71,** 1–10.

Amemiya, Y. and Anderson, T. W. (1985) *Asymptotic chi-square tests for a large class of factor analysis models,* Technical Report No. 13, Econometric Workshop, Stanford Univ.

Andersen, E. B. (1980) *Discrete Statistical Models with Social Science Applications.* Amsterdam: North-Holland Publishing Company.

Andersen, E. B. and Madsen, M. (1977) "Estimating parameters of the latent population distribution". *Psychometrika,* **42,** 357–74.

Anderson, J. C. and Gerbing, D. W. (1984) "The effect of sampling error on convergence, improper solutions, and goodness-of-fit indices for maximum likelihood confirmatory factor analysis". *Psychometrika,* **49,** 155–73.

Anderson, T. W. (1959) "Some scaling models and estimation procedures in the latent class model". In *Probability and Statistics, The Harald Cramér Volume,* ed. U. Grenander, Stockholm: Almqvist & Wicksell and Wiley, 9–38.

Anderson, T. W. (1984) *An Introduction to Multivariate Statistical Analysis,* 2nd edn. New York: Wiley.

Anderson, T. W. and Rubin, H. (1956) "Statistical inference in factor analysis". *Third Berkeley Symp. Math. Statist. and Prob.,* **5,** 111–50.

Barankin, E. W. and Maitra, E. P. (1963) "Generalisation of the Fisher–Darmois–Koopman–Pitman theorem on sufficient statistics". *Sankhyā, A,* **25,** 217–44.

Bartholomew, D. J. (1980) "Factor analysis for categorical data". *J. Roy. Statist. Soc.,* **B, 42,** 293–321.

Bartholomew, D. J. (1981a) *Mathematical Methods in Social Science,* Chichester: Wiley.

Bartholomew, D. J. (1981b) "Posterior analysis of the factor model". *Br. J. Math. Statist. Psychol.,* **34,** 93–9.

Bartholomew, D. J. (1983) "Latent variable models for ordered categorical data". *J. Econometrics,* **22,** 229–43.

Bartholomew, D. J. (1984a) "The foundations of factor analysis". *Biometrika,* **71,** 221–32.

Bartholomew, D. J. (1984b) "Scaling binary data using a factor model". *J. Roy. Statist. Soc.*, **B, 46,** 120–3.
Bartholomew, D. J. (1985a) "Foundations of factor analysis: Some practical implications". *Br. J. Math. Statist. Psychol.*, **38,** 1–10.
Bartholomew, D. J. (1985b) "Reply to discussion of foundations of factor analysis". *Br. J. Math. Statist. Psychol.*, **38,** 138–40.
Bartholomew, D. J. and McDonald R. P. (1986) "The foundations of factor analysis: a further comment". *Br. J. Math. Statist. Psychol.*, **39,** 228–9.
Bartlett, M. S. (1937) "The statistical conception of mental factors". *Br. J. Psychol.*, **28,** 97–104.
Bartlett, M. S. (1950) "Tests of significance in factor analysis". *Br. J. Psychol. (Statistical Sect.)*, **3,** 77–85.
Bartlett, M. S. (1953) "Factor analysis in psychology as a statistician sees it". *Uppsala Symp. Psychol. Factor Analysis*. Uppsala: Almqvist & Wiksell, 23–34.
Bartlett, M. S. (1954) "A note on the multiplying factors for various χ^2 approximations". *J. Roy. Statist. Soc.*, **B, 16,** 296–8.
Bearden, W. O., Sharma, S. and Teel, J. E. (1982) "Sample size effects on chi-square and other statistics used in evaluating causal models". *J. Market. Res.*, **19,** 425–30.
Bennett, N. (1976) *Teaching Styles and Pupil Progress*. London: Open Books.
Bentler, P. M. and Bonett, D. G. (1980) "Significance tests and goodness-of-fit in the analysis of covariance structures". *Psychol. Bull.*, **88,** 588–606.
Bentler, P. M. and Tanaka, J. S. (1983) "Problems with EM algorithms for ML factor analysis". *Psychometrika*, **48,** 247–51.
Berge, J. M. F. Ten (1983) "On Green's best linear composites with a specified structure, and oblique estimates of factor scores". *Psychometrika*, **48,** 371–5.
Bock, R. D. (1972) "Estimating item parameters and latent ability when responses are scored in two or more nominal categories". *Psychometrika*, **37,** 29–51.
Bock, R. D. and Aitkin, M. (1981) "Marginal maximum likelihood estimation of item parameters: application of an EM algorithm". *Psychometrika*, **46,** 443–59.
Bock, R. D. and Lieberman, M. (1970) "Fitting a response model for *n* dichotomously scored items". *Psychometrika*, **35,** 179–97.
Boomsma, A. (1985) "Nonconvergence, improper solutions, and starting values in LISREL maximum likelihood estimation". *Psychometrika*, **50,** 229–42.
Bozdogan, H. and Ramirez, D. E. (1986) *Model selection approach to the factor model problem, parameter parsimony, and choosing the number of factors*. Research Report, Dept of Mathematics, Univ. of Virginia, Charlottesville, Virginia.
Brokken, F. B. (1983) "Orthogonal procrustes rotation maximizing congruence". *Psychometrika*, **48,** 343–9.
Browne, M. W. (1969) "Fitting the factor analysis model". *Psychometrika*, **34,** 375–94.
Browne, M. W. (1974) "Generalized least squares estimators in the analysis of covariance structures". *S. Afr. J. Statist.*, **8,** 1–24. Reprinted in D. J. Aigner and A. S. Goldberg, (eds) (1977) *Latent Variables in Socio-economic Models*. Amsterdam: North-Holland.
Browne, M. W.(1982) "Covariance structures". In D. M. Hawkins (ed.) *Topics in Applied Multivariate Analysis*, 72–141. Cambridge: Univ. Press.
Browne, M. W. (1984) "Asymptotically distribution-free methods for the analysis of covariance structures". *Br. J. Math. Statist. Psychol.*, **37,** 62–83.
Burt, C. (1941) *The Factors of the Mind: An Introduction to Factor Analysis in Psychology*. New York: Macmillan.
Carroll, J. B. (1953) "An analytical solution for approximating simple structure in factor analysis". *Psychometrika*, **18,** 23–38.

Cattell, R. B. (1978) *The Scientific Use of Factor Analysis in Behavioral and Life Sciences*. New York: Plenum Press.

Chambers, R. G. (1982) "Correlation coefficients from 2 × 2 tables and from biserial data". *Br. J. Math. Statist. Psychol.*, **35,** 216–27.

Chatterjee, S. (1984) "Variance estimation in factor analysis: An application of the bootstrap". *Br. J. Math. Statist. Psychol.*, **37,** 252–62.

Christofferson, A. (1975) "Factor analysis of dichotomized variables". *Psychometrika*, **40,** 5–32.

Cliff, N. (1966) "Orthogonal rotation to congruence". *Psychometrika*, **31,** 33–42.

Cliff, N. (1983) "Some cautions concerning the application of causal modelling methods". *Multivariate Behavioral Res.*, **18,** 115–26.

Cliff, N. and Hamburger, C. D. (1967) "The study of sampling errors in factor analysis by means of artificial experiments". *Psychol. Bull.*, **68,** 430–45.

Cliff, N. and Pennell, R. (1967) "The influence of communality, factor strength, and loading size on the sampling characteristics of factor loadings. *Psychometrika*, **32,** 309–26.

Clogg, C. C. (1979) "Some latent structure models for the analysis of Likert-type data". *Social Sci. Res.*, **8,** 287–301.

Coleman, J. S. (1964) *Introduction to Mathematical Sociology*, New York: Free Press.

Comrey, A. L. (1962) "The minimum residual method of factor analysis". *Psychol. Reports*, **11,** 15–18.

Comrey, A. L. and Ahumada, A. (1964) "An improved procedure and program for minimum residual factor analysis". *Psychol. Reports*, **15,** 91–6.

Dempster, A. P., Laird, N. M. and Rubin, D. B. (1977) "Maximum likelihood from incomplete data via the EM algorithm". *J. Roy. Statist. Soc.*, **B, 39,** 1–38.

Derflinger, G. (1969) "Efficient methods for obtaining minres and maximum likelihood solutions in factor analysis". *Metrika*, **14,** 214–31.

Divgi, D. R. (1979) "Calculation of the tetrachoric correlation coefficient". *Psychometrika*, **44,** 169–72.

Dolby, G. R. (1976) "Structural relations and factor analysis", *The Math. Scientist*, Suppl. No. 1, 25–9.

Driel, O. P. van (1978) "On various causes of improper solutions in maximum likelihood factor analysis". *Psychometrika*, **43,** 225–43.

Duncan-Jones, P., Grayson, D. A. and Moran, P. A. P. (1986) "The utility of latent trait models in psychiatric epidemiology". *Psychol. Medicine*, **16,** 391–405.

Efron, B. (1982) "Transformation theory: How normal is a family of distributions?" *Ann. Statist.*, **10,** 323–39.

Efron, B. and Tibshirani, R. (1986) "Bootstrap methods for standard errors, confidence intervals and other measures of statistical accuracy". *Statist. Science*, **1,** 54–75.

Etezadi-Amoli, J. and McDonald, R. P. (1983) "A second generation nonlinear factor analysis". *Psychometrika*, **48,** 315–42.

Everitt, B. S. (1984) *An Introduction to Latent Variable Models*. London: Chapman & Hall.

Everitt, B. S. and Hand, D. J. (1985) *Finite Mixture Distributions*. London: Chapman & Hall.

Fachel, J. M. G. (1986) *The C-type Distribution as an Underlying Model for Categorical Data and its use in Factor Analysis*. Ph.D. Thesis, University of London.

Fischer, G. H. (1983) "Logistic latent trait models with linear constraints". *Psychometrika*, **48,** 3–26.

Francis, I. (1974) "Factor analysis: fact or fabrication". Invited address delivered at the

Seventh N. Z. Math. Colloquium, held at Christchurch, 8–10 May, 1972. *Math. Chronicle,* **3,** 9–44.

Fuller, E. L., Jr. and Hemmerle, W. J. (1966) "Robustness of the maximum-likelihood estimation procedure in factor analysis". *Psychometrika,* **31,** 255–66.

Gibson, W. A. (1959) "Three multivariate models: Factor analysis, latent structure analysis, and latent profile analysis". *Psychometrika,* **24,** 229–52.

Gibson, W. A. (1962) "Latent structure and positive manifold". *Br. J. Statist. Psychol.,* **15,** 149–60.

Goodman, L. A. (1978) *Analyzing Qualitative/Categorical Data* (ed. J. Magidson). Cambridge, Mass.: Abt Books.

Graybill, F. A. (1983), *Matrices with Applications in Statistics.* Belmont, California: Wadsworth International Group.

Green, B. F. (1952a) "The orthogonal approximation of an oblique structure in factor analysis". *Psychometrika,* **17,** 429–40.

Green, B. F. (1952b) "Latent structure analysis and its relation to factor analysis". *J. Am. Statist. Assoc.,* **47,** 71–6.

Green, B. F. (1969) "Best linear composites with a specific structure". *Psychometrika,* **34,** 301–18.

Greenacre, M. J. (1984) *Theory and Applications of Correspondence Analysis.* London: Academic Press.

Gustafsson, J.-E. (1980) "A solution of the conditional estimation problem for long tests in the Rasch model for dichotomous items". *Educ. and Psychol. Measurement,* **40,** 377–85.

Guttman, L. (1954) "A new approach to factor analysis: The Radex". Chapter in Paul F. Lazarsfeld (ed.) *Mathematical Thinking in the Social Sciences.* New York: Columbia Univ. Press, 258–348.

Guttman, L. (1955) "The determinacy of factor score matrices with implications for five other basic problems of common factor theory". *Br. J. Statist. Psychol.,* **8,** 65–82.

Guttman, L. (1968) What lies ahead for factor analysis? (Symposium: The future of factor analysis). *Educ. and Psychol. Measurement,* **18,** 497–515.

Gweke, J. F. and Singleton, K. J. (1980) "Interpreting the likelihood ratio statistic in factor models when sample size is small". *J. Am. Statist. Assoc.,* **75,** 133–7.

Hafner, R. (1981) "An improvement of the Harman–Fukuda method for the minres solution in factor analysis". *Psychometrika,* **46,** 347–9.

Hakstian, A. R., Rogers, W. T. and Cattell, R. B. (1982) "The behavior of number-of-factors rules with simulated data". *Multivariate Behavioral Res.,* **17,** 193–219.

Hambleton, R. and Cook, L. L. (1977) "Latent trait models and their use in the analysis of educational test data". *J. Educ. Measurement,* **14,** 75–96.

Harman, H. H. (1976) *Modern Factor Analysis,* 3rd edn. Chicago: Univ. of Chicago Press.

Harman, H. H. and Fukuda, Y. (1966) "Resolution of the Heywood case in the minres solution". *Psychometrika,* **31,** 563–71.

Harman, H. H. and Jones, W. H. (1966) "Factor analysis by minimizing residuals (Minres)". *Psychometrika,* **31,** 351–68.

Harris, C. W. and Kaiser, H. F. (1964) "Oblique factor analytic solutions by orthogonal transformations". *Psychometrika,* **29,** 347–62.

Healy, M. J. R. and Goldstein, H. (1976) "An approach to the scaling of categorized attributes". *Biometrika,* **63,** 219–29.

Hendrickson, A. E. and White, P. O. (1964) "Promax: A quick method for rotation to oblique simple structure". *Br. J. Statist. Psychol.,* **17,** 65–76.

Henrysson, S. (1950) "The significance of factor loading—Lawley's test examined by artificial samples". *Br. J. Psychol., Statistics Sect.*, **3,** 159–65.

Heywood, H. B. (1931) "On finite sequences of real numbers". *Proc. Roy. Soc.*, Ser. **A, 134,** 486–510.

Holland, P. W. (1981) "When are item response models consistent with observed data?" *Psychometrika,* **46,** 79–92.

Holzinger, K. J. and Swineford, F. (1939) "A study in factor analysis: the stability of a bi-factor solution". *Supplementary Educational Monographs,* No. 48. Chicago: Dept of Education, Univ. of Chicago.

Hope, A. C. A. (1968) "A simplified Monte Carlo significance test procedure". *J. Roy. Statist. Soc.*, **B, 30,** 582–98.

Hotelling, H. (1933) "Analysis of a complex of statistical variables into principal components". *J. Educ. Psychol.*, **24,** 417–41 and 498–520.

Howe, W. G. (1955) *Some Contributions to Factor Analysis.* Oak Ridge National Laboratory, Oak Ridge.

Humphreys, L. G., Ilgen, D., McGrath, D. and Montanelli, R. (1969) "Capitalizing on chance in rotation of factors". *Educ. and Psychol. Measurement,* **29,** 259–71.

Jackson, D. N. and Chan, D. W. (1980) "Maximum-likelihood estimation in common factor analysis: A cautionary note". *Psychol. Bull.*, **88,** 502–8.

Jackson, D. N. and Morf, M. E. (1973) "An empirical evaluation of factor reliability". *Multivariate Behavioral Res.*, **8,** 439–59.

Jansen, P. G. W. and Roskam, E. E. (1986) "Latent trait models and dichotomization of graded responses". *Psychometrika,* **51,** 69–91.

Jennrich, R. I. and Robinson, S. M. (1969) "A Newton–Raphson algorithm for maximum likelihood factor analysis". *Psychometrika,* **34,** 111–23.

Johnson, R. A. and Wichern, D. W. (1982) *Applied Multivariate Analysis.* Englewood Cliffs, N.J.: Prentice-Hall.

Jöreskog, K. G. (1963) *Statistical Estimation in Factor Analysis.* Stockholm: Almqvist & Wiksell.

Jöreskog, K. G. (1967) "Some contributions to maximum likelihood factor analysis". *Psychometrika,* **32,** 443–82.

Jöreskog, K. G. (1969) "A general approach to confirmatory maximum likelihood factor analysis". *Psychometrika,* **34,** 183–220.

Jöreskog, K. G. (1970) "A general method for analysis of covariance structures". *Biometrika,* **57,** 239–51.

Jöreskog, K. G. (1977) "Structural equation models in the social sciences: Specification, estimation and testing". In P. R. Krishnaiah (ed.), *Applications of Statistics,* Amsterdam: North-Holland.

Jöreskog, K. G. (1979) "Basic ideas of factor and component analysis". In K. G. Jöreskog and D. Sorbom (ed), *Advances in Factor Analysis and Structural Equation Models.* Cambridge, Mass.: Abt Books.

Jöreskog, K. G. and Goldberger, A. S. (1972) "Factor analysis by generalized least squares". *Psychometrika,* **37,** 243–59.

Jöreskog, K. G. and Sorbom, D. (1977) "Statistical models and methods for analysis of longitudinal data". In *Latent Variables in Socioeconomic Models,* ed. D. J. Aigner and A. S. Goldberger, Amsterdam: North-Holland, 285–325.

Kaiser, H. F. (1958) "The varimax criterion for analytic rotation in factor analysis". *Psychometrika,* **23,** 187–200.

Kano, Y. (1983) "Consistency of estimators in factor analysis". *J. Japan Statist. Soc.*, **13,** 137–44.

Kano, Y. (1984) "Construction of additional variables conforming to a common factor model". *Statist. and Prob. Letters,* **2,** 241–4.

Kano, Y. (1986a) "Conditions on consistency of estimators in covariance structure model". *J. Japan Statist. Soc.,* **16,** 75–80.

Kano, Y. (1986b) "Consistency conditions on the least squares estimator in a single common factor analysis model". *Annal. Inst. Statist. Math.,* **39, A,** 57–68.

Kendall, M. G. (1954) Review of Uppsala symposium on psychological factor analysis, *J. Roy. Statist. Soc.,* **A, 107,** 462–83.

Kendall, M. G. (1957, 1975) *A Course in Multivariate Analysis.* London: Griffin.

Kendall, M. G. and Babington Smith, B. (1950) "Factor analysis". *J. Roy. Statist. Soc.,* **B, 12,** 60–94.

Kendall, M. G. and Lawley, D. N. (1956) "The principles of factor analysis". *J. Roy. Statist. Soc.,* **A, 119,** 83–4.

Koopman, R. F. (1978) "On Bayesian estimation in unrestricted factor analysis". *Psychometrika,* **43,** 109–10.

Krane, W. R. and McDonald, R. P. (1978) "Scale invariance and the factor analysis of correlation matrices". *Br. J. Math. Statist. Psychol,* **31,** 218–28.

Lancaster, H. O. (1954) "Traces and cumulants of quadratic forms in normal variables". *J. Roy. Statist. Soc.,* **B, 16,** 247–54.

Lawley, D. N. and Maxwell, A. E.(1963, 2nd edn 1971) *Factor Analysis as a Statistical Method.* London: Butterworth.

Lawley, D. N. and Maxwell, A. E. (1973) "Regression and factor analysis", *Biometrika,* **60,** 331–8.

Lazarsfeld, P. F. and Henry, N. W. (1968) *Latent Structure Analysis.* New York: Houghton-Mifflin.

Lee, H. B. and Comrey, A. L. (1978) "An empirical comparison of two minimum residual factor extraction methods". *Multivariate Behavioral Res.,* **13,** 497–507.

Lee, S.-Y. (1981) "A Bayesian approach to confirmatory factor analysis". *Psychometrika,* **46,** 153–60.

Leimu, H. (1983). Työntekijäin työasema ja työpaikkaliikkuvuus erikokoisissa teollisuusyrityksissä. (Bluecollar work position and workplace mobility in small and large industrial firms, I. Theoretical points of departure and comparison of workers in small and large industrial firms). Univ. of Turku, C 40, Turku, 1983.

Lombard, H. L. and Doering, C. R. (1947). "Treatment of the fourfold table by partial association and partial correlation as it relates to public health problems". *Biometrics,* **3,** 123–8.

Lord, F. M. and Novick, M. R. (1968) *Statistical Theories of Mental Test Scores.* Reading, Mass.: Addison-Wesley.

Louis, T. A. (1982) "Finding the observed information matrix when using the EM algorithm". *J. Roy. Statist. Soc.,* **B, 44,** 226–33.

MacCallum, R. (1983) "A comparison of factor analysis programs in SPSS, BMDP and SAS". *Psychometrika,* **48,** 223–31.

McDonald, R. P. (1962a) "A general approach to non-linear factor analysis". *Psychometrika,* **27,** 397–415.

McDonald, R. P. (1962b) "A note on the derivation of the general latent class model". *Psychometrika,* **27,** 203–6.

McDonald, R. P. (1967a) "Non-linear factor analysis". *Psychometric Monographs,* No. 15.

McDonald, R. P. (1967b) "Numerical methods for polynomial models in non-linear factor analysis". *Psychometrika,* **32,** 77–112.

McDonald, R. P. (1969) "The common factor analysis of multicategory data". *Br. J. Math. Statist. Psychol.*, **22**, 165–75.
McDonald, R. P. (1974) "The measurement of factor indeterminacy". *Psychometrika*, **39**, 203–22.
McDonald, R. P. (1978) "A simple comprehensive model for the analysis of covariance structures". *Br. J. Math. Statist. Psychol.*, **31**, 59–72.
McDonald, R. P. (1979) "The simultaneous estimation of factor loadings and scores". *Br. J. Math. Statist. Psychol*, **32**, 212–28.
McDonald, R. P. (1985) *Factor Analysis and Related Methods*. New Jersey: Erlbaum Associates.
McDonald, R. P. and Burr, E. J. (1967) "A comparison of four methods of constructing factor scores". *Psychometrika*, **32**, 381–401.
McDonald, R. P. and Swaminathan, H. A. (1973) "A simple matrix calculus with applications to multivariate analysis". *General Systems*, **18**, 37–54.
McFadden, D. (1982) "Qualitative response models". *Advances in Econometrics*, W. Hildenbrand (ed.), Cambridge: Univ. Press, 1–37.
McHugh, R. B. (1956) "Efficient estimation and local identification in latent class analysis". *Psychometrika*, **21**, 331–47.
McHugh, R. B. (1958) "Note on 'Efficient estimation and local identification in latent class analysis'". *Psychometrika*, **23**, 273–4.
Macready, G. B. and Dayton, C. M. (1977) "The use of probabilistic models in the assessment of mastery". *J. Educ. Statist.*, **2**, 99–120.
Magnus, J. R. and Neudecker, H. (1986) *Matrix Differential Calculus and Static Optimization*. Chichester: Wiley.
Mardia, K. V. (1970), *Families of Bivariate Distributions*. London: Griffin.
Mardia, K. V., Kent, J. T. and Bibby, J. M. (1979) *Multivariate Analysis*. New York: Academic press.
Martin, J. K. and McDonald, R. P. (1975) "Bayesian estimation in unrestricted factor analysis; a treatment for Heywood cases". *Psychometrika*, **40**, 505–17.
Martinson, E. O. and Hadman, M. A. (1975). "Calculation of the polychoric estimate of correlation in contingency tables", Algorithm AS 87, *Applied Statist.*, **24**, 272–8.
Masters, G. M. (1985) "A comparison of latent trait and latent class analysis of Likert-type data". *Psychometrika*, **50**, 69–82.
Maxwell, A. E. (1961) "Recent trends in factor analysis". *J. Roy. Statist. Soc.*, **A, 124**, 49–59.
Mislevy, R. J. (1984) "Estimating latent distributions". *Psychometrika*, **49**, 359–81.
Mislevy, R. J. (1985). "Estimation of latent group effects". *J. Am. Statist. Assoc.*, **80**, 993–7.
Mooijaart, A. (1983) "Two kinds of factor analysis for ordered categorical variables". *Multivariate Behavioral Res.*, **18**, 423–41.
Mooijaart, A. (1985) "Factor analysis for non-normal variables". *Psychometrika*, **50**, 323–42.
Morrison, D. F. (1967) *Multivariate Statistical Methods*. New York: McGraw-Hill.
Mosteller, F. (1968) "Association and estimation in contingency tables". *J. Am. Statist. Assoc.*, **63**, 1–28.
Mulaik, S. A. (1972) *The Foundations of Factor Analysis*. New York: McGraw-Hill.
Mulaik, S. A. (1986). "Factor analysis and Psychometrika: major developments". *Psychometrika*, **51**, 23–33.
Mulaik, S. A. and McDonald, R. P. (1978) "The effect of additional variables on factor indeterminacy in models with a single common factor". *Psychometrika*, **43**, 177–92.

Muthén, B. (1978) "Contributions to factor analysis of dichotomous variables". *Psychometrika,* **43,** 551–60.
Muthén, B. and Christoffersson, A. (1981) "Simultaneous factor analysis of dichotomous variables in several groups". *Psychometrika,* **46,** 407–19.
Neuhaus, J. O. and Wrigley, C. (1954) "The quartimax method: An analytical approach to orthogonal simple structure". *Br. J. Statist. Psychol.,* **7,** 81–91.
Nosal, M. (1977) A note on the MINRES method. *Psychometrika,* **42,** 149–51.
Okamoto, M. and Ihara, M. (1983) "A new algorithm for the least-squares solution in factor analysis". *Psychometrika,* **48,** 597–605.
Olsson, U. (1979) "Maximum likelihood estimation of the polychoric coefficient". *Psychometrika,* **44,** 443–59.
Olsson, U., Drasgow, F. and Dorans, N. J. (1982) "The polyserial correlation coefficient". *Psychometrika,* **47,** 337–47.
Pearson, K. (1901) "On lines and planes of closest fit to a system of points in space". *Phil. Mag.,* **2,** 6th ser., 557–72.
Pearson, K. and Moul, M. (1927) "The mathematics of intelligence. I—The sampling errors in the theory of a generalized factor". *Biometrika,* **19,** 246–92.
Pennell, R. (1968) "The influence of communality and *N* on the sampling distribution of factor loadings". *Psychometrika,* **33,** 423–39.
Pennell, R. (1972) "Routinely computable confidence intervals for factor loadings using the 'jack-knife'". *Br. J. Math. Statist. Psychol.,* **25,** 107–14.
Pickering, R. M. and Forbes, J. F. (1984) "A classification of Scottish infants using latent class analysis". *Statist. in Medicine,* **3,** 249–59.
Plackett, R. L. (1965) "A class of bivariate distributions". *J. Am. Statist. Assoc.,* **60,** 516–22.
Rao, C. R. (1955) "Estimation and tests of significance in factor analysis". *Psychometrika,* **20,** 93–111.
Rasch, G. (1960) *Probabilistic models for some intelligence and attainment tests.* Copenhagen: Paedagogiske Institut.
Rigdon, S. E. and Tsutakawa, R. K. (1983) "Parameter estimation in latent trait models". *Psychometrika,* **48,** 567–74.
Rosenbaum, P. R. (1984) "Testing the conditional independence and monotonicity assumptions of item response theory". *Psychometrika,* **49,** 425–35.
Rubin, D. B. and Thayer, D. T. (1982) "EM algorithms for ML factor analysis". *Psychometrika,* **47,** 69–76.
Rubin, D. B. and Thayer, D. T. (1983) "More on EM for ML factor analysis". *Psychometrika,* **48,** 253–7.
Russell, M. A. H., Peto, J. and Patel, U. A. (1974) "The classification of smoking by factorial structure of motives". *J. Roy. Statist. Soc.,* **A, 137,** 313–46.
Samejima, F. (1974) "Normal ogive model on the continuous response level in the multi-dimensional latent space". *Psychometrika,* **39,** 111–21.
Sanathan, L. and Blumenthal, S. (1978) "The logistic model and estimation of latent structure". *J. Am. Statist. Assoc.,* **73,** 794–9.
Saunders, D. R. (1961) "The rationale for an 'oblimax' method of transformation in factor analysis". *Psychometrika,* **26,** 317–24.
Scheussler, K. F. (1982) *Measuring Social Life Feelings.* San Francisco: Jossey-Bass Inc.
Schönemann, P. H. A. (1966) "A generalized solution of the orthogonal procrustes problem". *Psychometrika,* **31,** 1–10.
Schönemann, P. H. A. (1981) "Power as a function of communality in factor analysis". *Bull. Psychonomic Soc.,* **17**(1), 57–60.

Schriever, B. F. (1983) "Scaling of order dependent categorical variables". *Int. Statist. Rev.*, **51**, 225–38.

Schwarz, G. (1978) "Estimating the dimension of a model", *Ann. Statist.*, **6**, 461–4.

Seber, G. A. F. (1984) *Multivariate Observations*, New York: Wiley.

Shapiro, A. (1985) "Asymptotic distribution of test statistics in the analysis of moment structures under inequality constraints". *Biometrika*, **72**, 133–44.

Shea, B. L. (1984) FACONE: A computer program for fitting the logit latent variable model by maximum likelihood. Dept of Statistics, London School of Economics.

Shea, B. L. (1985) POLYFAC: A computer program for fitting the logit latent variable model to ordered categorical data by maximum likelihood. Dept of Statistics, London School of Economics.

Smith, G. A. and Stanley, G. (1983) "Clocking *g*: relating intelligence and measures of timed performance". *Intelligence*, **7**, 353–68.

Spearman, C. (1904) "General intelligence, objectively determined and measured". *Am. J. Psychol.*, **15**, 201–93.

Spearman, C. (1927) *The Abilities of Man.* New York: Macmillan.

Stegelmann, W. (1983) "Expanding the Rasch model to a general model having more than one dimension". *Psychometrika*, **48**, 259–67.

Steiger, J. H. (1979) "Factor indeterminacy in the 1930's and the 1970's: Some interesting parallels". *Psychometrika*, **44**, 157–68.

Swain, A. J. (1975) "A class of factor analysis estimation procedures with common asymptotic sampling properties". *Psychometrika*, **40**, 315–35.

Tabachnick, B. G. and Fidell, L. S. (1983) *Using Multivariate Statistics*, New York: Harper & Row.

Tanaka, J. S. and Huba, G. J. (1985) "A fit index for covariance structure models under arbitrary GLS". *Br. J. Math. Statist. Psychol.*, **38**, 197–201.

Thissen, D. and Steinberg, L. (1984) "A response model for multiple choice items". *Psychometrika*, **49**, 501–19.

Thomson, G. H. (1939) *The Factorial Analysis of Human Ability*. University of London Press. (1951) 5th edn, New York: Houghton-Mifflin.

Thomson, G. H. (1946) *Some Recent Work in Factorial Analysis and a Retrospect*, Presidential address to British Psychol. Soc., London: Univ. of London Press.

Thurstone, L. L. (1935) *The Vectors of Mind.* Chicago: Univ. of Chicago Press.

Thurstone, L. L. (1947) *Multiple Factor Analysis*. Chicago: Univ. of Chicago Press.

Thurstone, L. L. and Thurstone, T. G. (1941) "Factorial studies of intelligence". *Psychometric Monographs*, No. 2.

Tucker, L. R., Koopman, R. F. and Linn, R. L. (1969) "Evaluation of factor analytic research procedures by means of simulated correlation matrices". *Psychometrika*, **34**, 421–59.

Vellicer, W. F., Peacock, A. C. and Jackson, D. N. (1982) "A comparison of component and factor patterns: A Monte Carlo approach". *Multivariate Behavioral Res.*, **17**, 371–88.

Wamani, W. T. (1985) *An Empirical Study of Latent Class Models: Application to Criterion-Referenced Tests*. Ph.D. thesis, Univ. of London.

Whittle, P. (1952) "On principal components and least square methods of factor analysis". *Skand. Akt.*, **35**, 223–39.

Williams, J. S. (1978) "A definition for the common-factor analysis model and the elimination of problems of factor score indeterminacy". *Psychometrika*, **43**, 293–306.

Williams, J. S. (1981) "A note on the uniqueness of minimum rank solutions in factor analysis". *Psychometrika,* **46,** 109–10.
Wu, C. F. J. (1983) "On the convergence of properties of the EM algorithm". *Ann. Statist.,* **11,** 95–103.
Zegers, F. E. and Berge, J. M. F. ten (1983) "A fast and simple computational method of minimum residual factor analysis". *Multivariate Behavioral Res.,* **18,** 331–40.

Index

Claudia and the Phantom Phone Calls